Machine Learning

Machine Learning: Theory and Practice provides an introduction to the most popular methods in machine learning. The book covers regression including regularization, tree-based methods including Random Forests and Boosted Trees, Artificial Neural Networks including Convolutional Neural Networks (CNNs), reinforcement learning, and unsupervised learning focused on clustering. Topics are introduced in a conceptual manner along with necessary mathematical details. The explanations are lucid, illustrated with figures and examples. For each machine learning method discussed, the book presents appropriate libraries in the R programming language along with programming examples.

Features:

- Provides an easy-to-read presentation of commonly used machine learning algorithms in a manner suitable for advanced undergraduate or beginning graduate students, and mathematically and/or programming-oriented individuals who want to learn machine learning on their own.
- Covers mathematical details of the machine learning algorithms discussed to ensure firm understanding, enabling further exploration.
- Presents worked out suitable programming examples, thus ensuring conceptual, theoretical and practical understanding of the machine learning methods.

This book is aimed primarily at introducing essential topics in Machine Learning to advanced undergraduates and beginning graduate students. The number of topics has been kept deliberately small so that it can all be covered in a semester or a quarter. The topics are covered in depth, within limits of what can be taught in a short period of time. Thus, the book can provide foundations that will empower a student to read advanced books and research papers.

Machine Learning
Theory and Practice

Jugal Kalita

CRC Press
Taylor & Francis Group
Boca Raton London New York

CRC Press is an imprint of the
Taylor & Francis Group, an **informa** business

A CHAPMAN & HALL BOOK

First edition published 2023

by CRC Press

6000 Broken Sound Parkway NW, Suite 300, Boca Raton, FL 33487-2742

and by CRC Press

4 Park Square, Milton Park, Abingdon, Oxon, OX14 4RN

CRC Press is an imprint of Taylor & Francis Group, LLC

Library of Congress Cataloging-in-Publication Data

Names: Kalita, Jugal Kumar, author.
Title: Machine learning : theory and practice / Jugal Kalita.
Description: First edition. | Boca Raton : Chapman & Hall/CRC Press, 2023.
| Includes bibliographical references and index. | Summary: "Machine
Learning: Theory and Practice provides an introduction to the most
popular methods in machine learning. The book covers regression
including regularization, tree-based methods including Random Forests
and Boosted Trees, Artificial Neural Networks including Convolutional
Neural Networks (CNNs), reinforcement learning, and unsupervised
learning focused on clustering. Topics are introduced in a conceptual
manner along with necessary mathematical details. The explanations are
lucid, illustrated with figures and examples. For each machine learning
method discussed, the book presents appropriate libraries in the R
programming language along with programming examples"-- Provided by publisher.
Identifiers: LCCN 2022032123 (print) | LCCN 2022032124 (ebook) | ISBN
9780367433543 (hardback) | ISBN 9780367433529 (paperback) | ISBN
9781003002611 (ebook)
Subjects: LCSH: Machine learning.
Classification: LCC Q325.5 .K35 2023 (print) | LCC Q325.5 (ebook) | DDC
006.3/1--dc23/eng20221102
LC record available at https://lccn.loc.gov/2022032123
LC ebook record available at https://lccn.loc.gov/2022032124

ISBN: 978-0-367-43354-3 (hbk)
ISBN: 978-0-367-43352-9 (pbk)
ISBN: 978-1-003-00261-1 (ebk)

DOI: 10.1201/9781003002611

Typeset in CMR10 font
by KnowledgeWorks Global Ltd.

Publisher's note: This book has been prepared from camera-ready copy provided by the authors.
Access the Support Material: www.routledge.com/9780367433543

*To my mother Nirala
and my father Benudhar*

Contents

Preface

The term machine learning covers the study of algorithms that can be used to build mathematical or computational models which can be created or learned from training data by inferring patterns. The learned models are then used in a predictive manner to perform tasks such as classification of previously unseen data. The goal in machine learning is to generalize from the training data so that the learned models work well with data on which the algorithm has not been trained.

Machine learning is a field of study by itself, quite closely related to data mining and data analytics. The area covers a lot of ground—mathematical as well as computational. In the recent past, machine learning has gained prominence due to the fact that it is finding applications in many disparate fields, such as computer vision to recognize faces and find objects, in natural language processing to translate from one language to another, in cybersecurity to identify intruding network traffic, and in healthcare to recognize patterns for gene expression that may lead to diseases.

This textbook introduces essential topics in machine learning to advanced undergraduates and beginning graduate students. The number of topics has been kept deliberately small so that it can all be covered in a semester or a quarter. The topics are covered in depth, within limits of what can be taught in a short period of time. Thus, the book can provide foundations that will empower a student to read advanced books and research papers.

The book covers topics in machine learning, both supervised as well unsupervised. The bulk of the book is dedicated to supervised learning. Supervised techniques covered include regression, decision-tree based classifiers including ensembles of trees, and neural networks including deep learning. All of these algorithms are useful in practice. Lately, deep learning has become synonymous with Artificial Intelligence as well as Machine Learning, and hence, this topic is covered in sufficient depth, but in a compact manner, within the scope of a single chapter. Unsupervised learning is covered in one chapter only and covers clustering. The book also covers the important topic of reinforcement learning, where the objective for an agent is to learn how to choose a sequence of most beneficial actions to perform, in pursuit of a long-term goal. The agent learns to do so from rare rewards and punishments, received episodically.

There are quite a few textbooks in the market on machine learning. Most of these books are advanced, requiring heavy knowledge of mathematics for which many students are unprepared. On the other hand, there are many other "trade" books that cover the theoretical concepts in a scanty manner, but focus on only the programming aspects in a particular programming language of choice. This book is somewhere in between. It covers theory as well as practice. The theory is not covered with as much math as in some books, but in a sufficient amount of detail for a solid understanding of the topics.

Knowing the concepts and the theory behind a machine learning algorithm or approach allows the student to apply the techniques to various problems as a practitioner of the field. We provide ample examples so that the student can understand the essence of an algorithm or approach. We include enough mathematical details, starting from the simple to more complex, as needed. The advanced mathematical concepts will open doors to students who are interested in performing research in machine learning.

To understand the machine learning techniques, one needs to know how to program the algorithms from scratch, but to be able to use them to solve real life problems, the student needs to learn how to use languages and tools that are more suitable for the tasks at hand. These days, the predominant programming languages used in academia as well as industry, are Python and R. R is a statistical package, with an associated language that has powerful libraries for all kinds of machine learning tasks. Python is a general purpose language that has become an important vehicle for prototyping machine learning algorithms. It also has libraries that support a great number of machine learning algorithms. We will also discuss specific libraries like TensorFlow and Keras in the context of deep learning. We include examples in R in the current edition of the book. The programs discussed in the book will be available in a companion website for download for the benefit of the reader.

About the Author

Dr. Jugal Kalita teaches Computer Science at the University of Colorado, Colorado Springs, where he has been a professor since 1990. He received M.S. and Ph.D. degrees in Computer and Information Science from the University of Pennsylvania in Philadelphia in 1988 and 1990, respectively. Prior to that, he had received an M.Sc. in Computational Science from the University of Saskatchewan in Saskatoon, Canada in 1984; and a B.Tech. in Computer Science and Engineering from the Indian Institute of Technology, Kharagpur in 1982.

Dr. Jugal Kalita's expertise is in the areas of Artificial Intelligence and Machine Learning, and the application of techniques in Machine Learning to Natural Language Processing, Network Security, and Bioinformatics. At the University of Colorado, Colorado Springs, and Tezpur University, Assam, India, where he is an adjunct professor, Dr. Kalita has supervised 15 Ph.D. and 125 M.S. students to graduation, and has mentored 100 undergraduates in independent research. He has published 250 papers in journals and refereed conferences, including prestigious conferences such as International Conference on Machine Learning (ICML), Association for Advancement of Artificial Intelligence (AAAI), North American Chapter of the Association for Computational Linguistics (NAACL), International Conference on Computational Linguistics (COLING) and Empirical Methods in Natural Language Processing (EMNLP). Dr. Kalita is the author of On Perl: Perl for Students and Professionals, Universal Press, 2003. He is also a co-author of Network Anomaly Detection: A Machine Learning Perspective, CRC Press, 2013; DDOS Attacks: Evolution, Detection, Prevention, Reaction and Tolerance, CRC Press, 2016; Network Traffic Anomaly Detection and Prevention: Concepts, Techniques, and Tools, Springer Nature, 2017; and Gene Expression Data Analysis, A Statistical and Machine Learning Perspective, CRC Press, 2021.

Dr. Kalita has received several teaching, research and service awards at the University of Colorado, Colorado Springs, in the Department of Computer Science, and the College of Engineering and Applied Science. He received the prestigious Chancellor's Award at the University of Colorado, Colorado Springs, in 2011, in recognition of lifelong excellence in teaching, research and service. More details about Dr. Kalita can be found at http://www.cs.uccs.edu/~kalita.

Chapter 1

Introduction

The world we live in is inundated with data from innumerable sources. Devices connected to the net produced sixty-four zettabytes (64 followed by 21 zeroes) bytes of data every day in 2020[1]. Amazon sold 4000 items per minute in its US marketplace for a total of 2.1 billion items in 2019[2]. Large amounts of data are created by other big stores—web-based as well physical—such as Walmart, Ebay and BestBuy as well. As of early 2021, half a million tweets were posted, half a million Facebook comments written, four million YouTube videos viewed, and two and a half million searches performed on Google every minute of every day[3]. Scientists are well on their way to sequencing millions human genomes, with each genome containing approximately three billion base pairs.

The point being made here is that so much data are being generated by organizations, large and small, for-profit and non-profit, and governmental or non-governmental, that it is impossible to make sense of even a tiny fraction of the data. Analysis of the data to provide beneficial services and/or generate higher profits requires sophisticated computer algorithms. Such computational approaches attempt to find nuggets of useful information in very large amounts of data. The dramatic examples given above pertain to some of the biggest organizations in the world. However, even much smaller organizations also generate and have to deal with astounding amounts of data, in quantities unimaginable even a few years ago. These days, most organizations—companies, non-profits, universities and governments—big, medium and even small—have to collect and mine useful patterns in large amounts of data to make decisions and predictions, to survive, grow and thrive. This involves automatically learning from data, or in other words, machine learning.

---

[1]https://datacenternews.asia/story/the-world-generated-64-2-zettabytes-of-data-last-year-but-where-did-it-all-go

[2]marketplace.com

[3]bernardmarr.com

DOI: 10.1201/9781003002611-1

1.1 Learning

Machine learning is a field of study that is making large impacts on the world awash with data. However, let us first briefly introduce the term *learning* in the context of humans before delving into machine learning, the topic of this book. There are many definitions of learning although we present just a couple. Brown, Roediger and McDaniel define learning as [5]

> *acquiring knowledge and skills and having them readily available from memory so you can make sense of future problems and opportunities.*

According to this definition, the process of learning involves acquiring new knowledge or skills and creating representation for them in human memory. Availability in memory makes the new knowledge readily accessible for future use by an agent. As more new knowledge elements or skills are learned, the corresponding representations are stored and indexed for quick retrieval and use.

Another definition of learning by Gross states that [27]

> *learning is the act of acquiring new, or modifying and reinforcing existing knowledge, behaviors, skills, values, or preferences which may lead to a potential change in synthesizing information, depth or the knowledge, attitude or behavior relative to the type and range of experience.*

This definition extends the previous one by referring to modification of existing representations. It also extends the target of what is learned to add values and preferences, besides knowledge and skills. It also adds that "to make sense of future problems and opportunities", the process of learning must synthesize what is learned with what is already known, and presumably this happens in memory.

1.2 Machine Learning

Machine learning pertains to learning by machines or computers instead of by humans. In general, the concept of learning is the same whether the learner is human or machine. Although machine learning can potentially be general, the current state of the art in machine learning is usually limited to dealing with single individual problems. For example, in computer vision, there is great emphasis on image classification—where a program learns how to recognize the type of an object—e.g., whether a nicely taken and centered photographic image is that of a dog or a cat or a car. While being able to tell the type of an object is an important skill to have, it is difficult to claim that such an ability makes an agent artificially intelligent in any general sense. It is a very simple ingredient of general artificial intelligence, and hundreds, if not thousands, of such skills may be necessary for an agent to claim intelligence. In addition, these skills should be available in a prompt and integrated—holistic fashion for general intelligence.

Machine learning pertains to learning to perform a task based on prior experience or from data. The task may be classifying various types of animal pictures, recognizing hand-written digits, being able to tell one face from another, or being able to drive a car autonomously. To learn how to recognize one hand-written digit (say, "1") from another (say, "2"), the machine learning program needs access to possibly thousands of examples of hand-written 1s and 2s. Similarly, to be able to learn how to tell each of ten digits apart with high accuracy, the learning program needs a dataset with thousands of examples of such digits. In fact, there is a widely available dataset called MNIST that can be used to train and test machine learning programs. When we seek to build a machine learning program for this task of hand-written digit recognition, it is necessary to measure how good its learning is. A simple way to measure performance may be accuracy. This metric allows us to measure performance of a machine learning program as it is trained on various amounts and/or qualities of data. The use of a metric also allows us to compare different machine learning programs.

Below are definitions of machine learning from two very well-known scientists. Samuel, who wrote the first checkers-playing program, defined machine learning in 1959 as a [29]

> *field of study that gives computers the ability to learn without being explicitly programmed.*

Samuel's definition, being a very early one, is quite generic, and does not refer to specific tasks or any way of measuring learning. A more elaborate second definition by Mitchell from 1997 is as follows [20]:

> *A computer program R is said to learn from experience E with respect to task T and performance measure P, if its performance at task T as measured by P, improves with experience.*

This definition is much more detailed and specific. For example, task T can be classification of hand-written digits or images of various things or objects. The performance measure P may be accuracy that measures the percentage of examples in a test set that are classified correctly. The experience E in this case is that the learning agent sees the training examples one by one. As it trains on more data from the MNIST or a similar dataset, the performance is likely to improve.

Another example task T may be how to play chess. We may want to write a program to play chess using what is called reinforcement learning, in contrast to classification. The experience E may refer to the learning program playing with a clone of itself. As it plays more and more, it becomes a more proficient player. In such a case, a possible performance measure may be the percentage of times one of the clones wins games against a human player (or another well-established software player) with a certain desired rating, assuming the games are played after every 10000 (say) iterations of co-evolution.

1.3   Types of Machine Learning

Depending on the type of data used and the approach taken to learn, there are several general types of machine learning: supervised, unsupervised and reinforcement learning. Arguably, the most popular form of machine learning is supervised learning. Unsupervised learning is usually used for exploratory analysis of data when not much is known. Reinforcement learning is used when an agent wants to learn how to perform a task as a sequence of actions, in a certain environment based on limited input from the environment or from a teacher or an observer.

1.3.1   Supervised Learning

In supervised learning, the datasets used are labeled. In other words, each example in the dataset has an associated "resulting" or "summary" value that depends on the specifics of the example. This is in addition to the attributes or features used to describe the details of the example. There are two types of supervised learning: classification and regression.

1.3.1.1   Classification

Let us consider the MNIST dataset. Each example has a black-and-white (or, grayscale) image which is 28×28 in size, with a label. In other words, each image is described in terms of 784 pixel values, each value between 0 and 255. MNIST has a total of 70,000 images of which 60,000 are used to train a machine learning model, and 10,000 are used to test the trained model. A model simply means a machine learning program, which is software implementation of a machine learning algorithm, possibly well-known. Each image also has a label in the sense that someone has attached an annotation saying the image represents the actual digit "1" or digit "9", or so and so forth. Such an attached note is called a *label*. In machine learning parlance, the label indicates the *class* the image belongs to. Thus, in the MNIST dataset, each image belongs to one of 10 possible classes. Machine learning algorithms use the 60,000 training set of images to learn a model or internal description of the different classes. The learned model can discriminate or differentiate among the ten classes as a result of the learning process. To test a learned model, the model is asked to classify the images for which the label is not known.

Another example of supervised learning is classifying flowers on the basis of measurements of aspects of their physical size. There is a well-known dataset called the Iris dataset, which is all about iris flowers only. It is a much simpler dataset with only 150 examples or rows. Unlike the MNIST dataset, where each of the 784 features are undifferentiated pixel values in 784 different locations in the image, in the Iris dataset—each example is described in terms of four attributes or features. The features are all numeric. They are sepal length, sepal width, petal length and petal width. Each example has a label called *species*. The dataset is not broken up into training and testing subsets. Although very simple, the dataset is often used to train and test machine learning algorithms or models, especially in academic settings. For example, we can randomly pick a large percentage (say, 80%) of the examples to train a classifier, and use the rest 20% for testing. We can report results in terms of how many of the test examples are classified correctly. Better still, we can run the experiment—which consists of random picking of a training subset, testing on the rest, obtaining accuracy—several times, and report the average results, possibly with standard deviation.

1.3.1.2 Regression

In regression, the label associated with an example in the dataset is numeric, rather than discrete. In classification discussed earlier, the labels are discrete. For example, the MNIST dataset has ten discrete labels or classes, and the Iris dataset has three distinct labels. In contrast, consider a dataset where the average dollar incomes of groups of immigrants in the US are given in terms of three numeric attributes or features—percentage speaking English, percentage of literacy and percentage in the US for five or more years. The dataset has only 35 examples or rows. The goal in this case may be to predict the income of a nationality group not listed in the dataset, given the characteristics.

Another example of regression may be learning to predict the angle at which a car's steering wheel should be turned next at a particular moment in time. In this case, the training examples may come from a dataset that contains images of the lane on the road directly in front of a car. In a simple dataset, the road image may contain just the lane in front such that it is always empty although there may be curvatures and other details that show condition of the road such as potholes or the presence of snow or water. In such a dataset, the ideal angle by which the steering wheel should be turned is also given. Given a number of such images along with the labels (which is the degree of turn of the steering wheel), a machine learning algorithm may be able to learn how much to turn the steering wheel given an unseen picture of the road immediately before it. Of course, the dataset described so far is simplistic. A realistic dataset is likely to have not only the picture of the empty lane ahead, but also pictures taken from other cameras in a car—such as other lanes in front, the side of the road, the presence of dividers and other markers on the road, the lanes behind, signs in front, above and side of the road, and so on and so forth. In a real self-driving car, the dataset is likely to be in the form of a video feed from several cameras and values obtained from a number of sensors placed in various positions in the car. In this case, the label may come from an expert human driver actually driving a car. The goal would be to collect detailed data from an expert driver and use the data to learn to turn the steering wheel automatically like an expert!

1.4　Unsupervised Machine Learning or Clustering

In supervised learning, we perform classification or regression using discreet or continuous labels, respectively, assigned to examples in the dataset. Since the labels are given by an "outside individual" or "teacher", such methods are called supervised learning methods. In contrast, if the dataset is not labeled, and we perform machine learning on such a dataset, it is called *unsupervised learning*. Like supervised learning, unsupervised learning also comes in many forms such as clustering, outlier mining and association rule mining. In this book, we discuss clustering only, since it is arguably the most commonly used form of unsupervised learning.

For an academic discussion, we can convert a labeled dataset used to perform classification into an unlabeled dataset by removing the labels. For example, we can simply remove the species column (label) in the Iris dataset to obtain and unlabeled Iris dataset with 3 columns and 150 rows. Similarly, we can remove the digit column (label) in the MNIST dataset to obtain a dataset of $70,000$ examples, where each example consists of 784 pixels corresponding to hand-written digits.

Given such an unlabeled dataset, we may want to obtain "natural" groupings or clusters in the data. We may ask an unsupervised machine learning system to group the unlabeled MNIST dataset or the unlabeled Iris dataset into K groups where K is a small positive integer. The value of K does not have to be equal to the number of classes in clustering; it can potentially be any small positive integer.

To cluster a set of unlabeled data examples, we need a way to compute distance between two arbitrary examples. Assuming each attribute or feature of the example is numeric and continuous, we can use Euclidean distance. Given this distance metric, the objective of clustering is to separate a given dataset into a set of disjoint subsets such that examples in a subset are as similar (close in distance) as possible, and examples in different subsets are as dissimilar or distant as possible.

A dataset to cluster does not have to be the unlabeled version of a labeled dataset. Clustering is usually used for preliminary exploration of a new dataset to get insights into it. Some clustering algorithms like K-means have to know the number of clusters to look for to start, whereas other clustering algorithms such as hierarchical and density-based clustering can automatically determine the optimal number of clusters.

1.5 Reinforcement Learning

Supervised learning needs input from a teacher or supervisor regarding the classes data examples belong to, and learns to recognize the class of an unseen example. In contrast, unsupervised learning is given an unlabeled (or unclassed or uncategorized) dataset and is asked to find natural groupings (or clusters) in the data. Although the terms classes and clusters seem similar, the term *class* is used in supervised learning whereas the term *cluster* is used in unsupervised learning—the two terms should not be confused.

Reinforcement learning is neither supervised nor unsupervised. Some people think of it as unlike either, whereas others say it has characteristics of both. In reinforcement learning, an agent learns to perform a task within an environment. An example task can be learning to play checkers or chess, or to traverse a maze, or to autonomously drive a car. A reinforcement learning agent has a repertoire or a set of basic actions it can perform, and at any moment is assumed to "reside" in one of a set of states. For example, in learning to go through a maze, the action set may consist of going left, going right, going up and going down in a rectangular grid representing the maze. The maze may be strewn with obstacles and not all actions may be possible to be performed in all states. In learning to traverse a maze, the agent performs a sequence of actions that leads the agent from a source state to a goal state in the environment. The state of an agent can be thought of as the agent's position in a two-dimensional grid. When the agent reaches the goal state, the environment, a teacher or the agent itself gives a reward. Most actions are unrewarded, but rewards are given on an infrequent or "rare" manner. For example, in the case of the maze learning, the mundane actions such as moving in different directions are usually unrewarded (or, get reward of 0), whereas the last action that takes the agent out of the maze gets a high positive reward, say +100. Based on infrequent rewards as outlined above, the agent needs to learn what action is best to perform in which state to optimally perform the overall task like traversing a maze. Such a mapping of states to actions is called a *policy*. Thus, one of the main goals of reinforcement learning is to learn an optimal policy for a certain task.

In learning how to play chess also, the agent can be thought of as residing in one of a number of states during the game. In this case, a state of the agent can be the same as the configuration of the chess board, and thus fairly complex, requiring the position of each current piece on an 8×8 board. The possible action set would be the moves the agent can perform on the pieces. The reward structure depends on the program designer. For example, we can give a big positive immediate reward (say, $+100$) when the agent wins the game, and a big negative immediate reward (say, -100) when the agent loses. We can keep all other immediate rewards as 0. Alternatively, we can give some small positive or negative immediate rewards for taking pieces from the opponent or when the opponent takes the agent's pieces, obeying the rules of chess. Based on the repertoire of actions, the description of the states and the receipt of infrequent positive or negative immediate rewards, the agent needs to learn an optimal winning policy—which move to make in which state in any game of chess so that the agent may win.

1.6   Organization of the Book

This book is geared toward undergraduates and beginning graduate students in academic institutions around the world. It is also suitable for self-learners, whether students in academic settings, professionals in software-related industries, or a lay person with some background in college level mathematics and programming.

The book covers enough material for a semester of a first course on machine learning. It covers the basics of machine learning to prepare a student to understand more advanced material either through classes that follow, or through further self-reading and practice.

The book starts with a chapter on regression. Regression, arguably, forms the conceptual basis of supervised machine learning. A supervised machine learning algorithm takes labeled inputs to train on and produces a learned model that is able to predict labels (or values) for previously unseen inputs. The predicted value or label is sometimes numerical and sometimes not. If it is numerical, the approach is regression. However, even if the predicted value is discrete—the so-called problem of classification—many modern classifiers do not predict classes with certainty, but provide numeric values, which are probabilities of the unseen example belonging to the predicted classes. Thus, even classification can be thought of as a variant of regression. The second chapter of the book introduces regression to provide a good basis for supervised machine learning. It also introduces advanced topics on regularization of regression.

Chapter 3 focuses on classification using tree-based techniques. Decision tress and ensembles of decision trees in the form of random forests and boosting trees are strong classifiers, and are used in many practical problems in machine learning. Decision trees are also easy to interpret although it becomes somewhat difficult when a large number of trees make decisions in a collective or an ensemble.

Chapter 4 presents artificial neural networks. Artificial neural networks have been studied from the 1940s although they have become a dominant approach to machine learning in the past decade or so. Neural networks, traditionally, have been dense or fully-connected and shallow with just a few layers of nodes. However, the availability of economical graphical processing units (GPUs) and large amounts of trainable data in the recent past have led to the development and practical use of many-layered neural networks to solve challenging problems in machine learning better than ever, leading to frenzied excitement in this area, rechristened as *deep learning*. This chapter covers basics of artificial neural networks, how they learn as well as advanced topics such as convolutional neural networks (CNNs).

Chapter 5 covers the topic of reinforcement learning, introduced earlier. Reinforcement learning is about learning how to perform tasks by performing a sequence of basic actions to achieve a goal as optimally as possible. Learning, in this case, is guided by the need to obtain the highest possible cumulative rewards obtainable by performing a task, although individual immediate rewards, which are positive or negative, are "rarely" given to or obtained by the learning agent. Based on infrequent rewards, the learning agent reflects or contemplates to figure out which action in which state or circumstance led to receiving of optimal cumulative rewards. Learning which action to perform in which state to obtain most cumulative rewards in general is called learning a *policy*. This chapter covers the fundamental concepts in reinforcement learning.

Chapter 6 is on unsupervised machine learning, in particular, clustering. As noted earlier, in supervised learning, the data items are labeled. This means that some individual(s) may have gone through each data example and labeled it. Even if an existing program may have performed the initial labeling, one or more humans are likely to have checked the accuracy of the labels. Clustering is more of an exploratory or preparatory approach to machine learning, working with unlabeled data. It attempts to obtain natural groupings of unlabeled data, given a way to compute distance (or, conversely similarity) between two individual examples, between an individual example and a set of examples, or between two disjoint sets of examples. In this chapter, we discuss basic clustering techniques such as K-means that require the number of clusters K to be given to the clustering program. The chapter also covers hierarchical clustering and the more advanced density-based clustering—approaches that do not require K to be known before starting the clustering process.

Chapter 7 concludes it. This chapter provides a brief discussion of resources for a student who wants to delve deeper into machine learning, including deep learning.

1.7 Programming Language Used

Machine learning is of great practical use. Thus, learning the concepts, theory and algorithms is not enough for most learners. One has to learn how to program machine learning solutions for practical problems. This book not only introduces and explains the concepts, mathematics and insights behind machine learning algorithms, but also discusses how software libraries can be used effectively. There are powerful machine learning libraries in languages such as Python, R, Matlab and Java. Python is very popular these days due to the availability of deep learning packages such as TensorFlow, Keras and PyTorch, as well as traditional machine learning packages such as *scikit-learn*. R is also a popular language in data analytics and machine learning, with access to powerful machine learning libraries. Most books and websites are focused on using Python for machine learning. That is why when writing this book, we decided to use R as the language for the examples. R is popular in math, statistics, bioinformatics, computational medicine and many other fields. Although the examples are in R in the book, we intend to provide parallel examples in Python in a companion website.

1.8 Summary

This book provides strong conceptual and practical foundations in machine learning. This book will prepare the student adequately to pursue advanced topics in machine learning. Although deep learning is the topic of the day, it is necessary to learn the fundamentals to have a solid understanding of machine learning. This book provides these foundations. The book has a companion website where all corrections and updates, as well as accompanying code in R and Python will be provided.

Chapter 2

Regression

Often we are given a dataset for supervised learning, where each example in the training set is described in terms of a number of features, and the label associated with the example is numeric. In terms of mathematics, we can think of the features as independent variables and the label as a dependent variable. For example, the independent variable can be the monthly income of a person, and the dependent variable can be the amount of money the person spends on entertainment per month. In this case, we can also say that the person is described in terms of one feature, the income; and the person's label is the amount of money spent on entertainment per month. In such a case, our training set will consist of a number of examples where each person is described only in terms of his or her monthly income. Corresponding to each person, we have a label, which corresponds to the person's entertainment expense per month. Usually, the example is written as x, and the label as y. If we have N examples in our training set, we can refer to the ith example as $x^{(i)}$ and its corresponding label as $y^{(i)}$. The goal in regression is to find a function $\hat{f}$ of x that explains the training data the best. The ˆ on top of f says that it is not the real function f that explains the relationship between y and x, but an empirical approximation of it, based on the few data points we have been given. In machine learning, this function $\hat{f}$ is learned from the given dataset, so that it can be used to predict the value y, given an arbitrary value x' of x, i.e., $\hat{f}(x')$ is the predicted value $\hat{y}$ for y for a value x' for x. The dataset from which the regression function is learned is called the training dataset.

Of course, we can make the training examples more complex, if we describe each person in terms of several features, viz., the person's age, income per month, number of years of education, and the amount of money in the bank. Assume the label is still the same. In this case, each example is a vector of four features. In general, the example is a vector $\mathbf{x}$ (also written as $\vec{x}$) where each component is a specific feature of the example, and the label is still called y. If we have a training set with a number of examples, the ith example can be referred to as $\mathbf{x}^{(i)}$ and the corresponding label is $y^{(i)}$. The example $\mathbf{x}^{(i)}$ is described in terms of its features. If we have n features, the ith example can be written as

$$\mathbf{x}^{(i)} = [x_1^{(i)}, \cdots x_j^{(i)} \cdots x_n^{(i)}]$$

DOI: 10.1201/9781003002611-2

TABLE 2.1: LSD level in tissue and performance on a Math exam (LSDMath Dataset)

LSD	Math
78.93	1.17
58.20	2.97
67.47	3.26
37.47	4.69
45.65	5.83
32.92	6.00
29.97	6.41

where $x_j^{(i)}$ is the jth feature's value for the ith example. In performing regression, the goal is to learn a function $\hat{f}(\mathbf{x}')$ that predicts the value y, given an arbitrary vector value $\mathbf{x}'$.

Table 2.1 shows a dataset obtained from a University of Florida website[1] that shows the concentration of LSD in a student's tissue and the score on a math exam. The data is based on [33]. The independent variable is LSD concentration and the dependent variable is the score. This is an old dataset from the 1960s when the use of LSD was supposedly fairly common on US college campuses. The goal is to learn a function that predicts an unseen student's math score given his/her value of LSD concentration.

Table 2.2 is a second real dataset from 1909, obtained from the same University of Florida website, based on paper by [15]. Each row corresponds to an immigrant nationality group such as Irish, Turkish, Dutch or Slovak. The independent variables are the percentage of an immigrant group that spoke English, the percentage that was literate, and the percentage that had been living in the US for more than 5 years. The dependent variable is the average weekly wage in dollars. For purposes of discussion, we ignore the first column that spells out the nationality of an immigrant group. In other words, assume only each nationality group is described by three features, and the label or dependent variable is Income. The goal in performing regression will be to learn a function that can predict an arbitrary immigrant group's average income, given the percentage that spoke English, the percentage that was literate, and the percentage living in the US for more than five years.

We will first work with sample datasets with one independent variable and a corresponding dependent variable. We will then work with more complex datasets with several independent variables and a dependent variable. In this chapter, we will learn a way to find such a predictive function $\hat{f}$ using an approach called regression.

---

[1]http://www.stat.ufl.edu/~winner/datasets.html

TABLE 2.2: Shows Percentage speaking English, Percentage literate, Percentage in the US for 5 or more years, and Income for Immigrant groups in the US in 1909.

Nationality	English	Literacy	Residency	Income
Armenian	54.9	92.1	54.6	9.73
Bohemian/Moravian	66.0	96.8	71.2	13.07
Bulgarian	20.3	78.2	8.5	10.31
Canadian (French)	79.4	84.1	86.7	10.62
Canadian (Other)	100.0	99.0	90.8	14.15
Croatian	50.9	70.7	38.9	11.37
Danish	96.5	99.2	85.4	14.32
Dutch	86.1	97.9	81.9	12.04
English	100.0	98.9	80.6	14.13
Finnish	50.3	99.1	53.6	13.27
Flemish	45.6	92.1	32.9	11.07
French	68.6	94.3	70.1	12.92
German	87.5	98.0	86.4	13.63
Greek	33.5	84.2	18.0	8.41
Hebrew (Russian)	74.7	93.3	57.1	12.71
Hebrew (Other)	79.5	92.8	73.8	14.37
Irish	100.0	96.0	90.6	13.01
Italian (Northern)	58.8	85.0	55.2	11.28
Italian (Southern)	48.7	69.3	47.8	9.61
Lithuanian	51.3	78.5	53.8	11.03
Macedonian	21.1	69.4	2.0	8.95
Magyar	46.4	90.9	44.1	11.65
Norwegian	96.9	99.7	79.3	15.28
Polish	43.5	80.1	54.1	11.06
Portuguese	45.2	47.8	57.5	8.10
Roumanian	33.3	83.3	12.0	10.90
Russian	43.6	74.6	38.0	11.01
Ruthenian	36.8	65.9	39.6	9.92
Scotch	100.0	99.6	83.6	15.24
Servian	41.2	71.5	31.4	10.75
Slovak	55.6	84.5	60.0	11.95
Slovenian	51.7	87.3	49.9	12.15
Swedish	94.7	99.8	87.4	15.36
Syrian	54.6	75.1	45.3	8.12
Turkish	22.5	56.5	10.0	7.65

2.1 Regression

In regression, given a training dataset like the ones discussed above, a machine learning program learns a function or a model that describes the trend in the value of the dependent variable (or, the label) in terms of the values of

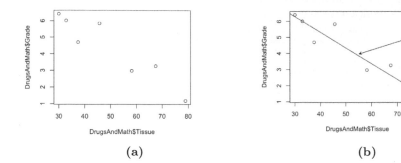

FIGURE 2.1: Scatter Plot and Linear Regression Line for the Drugs and Grades dataset

the independent variables, i.e., the feature values. As already discussed, the goal is to be able to predict the value of the dependent variable when a new unseen example is presented to the learned model. The regression function or model learned should be such that it generalizes well from the data, which is likely to have errors of observation as well as transcription. In other words, the model learned should not overfit to the training data, i.e., it should not just remember the training data, but should learn the essence of the trend in the data so that it can predict values for unseen examples, and do so well without making too much error.

2.1.1   Linear Least Squares Regression

Consider the first dataset given in Table 2.1. Suppose, we want to fit a straight line that describes the relationship between the amount of drug in the tissue of an exam taker and the score he or she receives on the test. A scatter plot for the dataset is shown in Figure 2.1(a). By eyeballing the scatter plot, we see that there is potentially a straight-line relationship going from the top left corner to the bottom right corner, although it may not be a great fit from the looks of it. A regression line we draw may look like the one shown in Figure 2.1(b). However, there are many possible lines we can draw as the linear fit to the data. Which line should we draw?

Visually, the training dataset can be seen like the matrix or table as given in Table 2.3. Note that the first column is for illustration only; it is not usually there in the data table, and even if it is there, we will ignore it. We assume there are N data examples.

We want to fit a line defined by the equation $y = \theta_1 x + \theta_0$ to the data points given, assuming each data example is specified as a single scalar independent variable x. The assumption in regression or any machine learning situation is that the data are not perfect, and that the data have flaws in them due to reasons such as observation errors, record-keeping errors and transcription

TABLE 2.3: General Data Matrix

No	$x_1 \cdots x_n$	y
1	$\cdots$	.
2	$\cdots$	.
$\vdots$	$\vdots$	$\vdots$
i	$x_1^{(i)} \cdots x_n^{(i)}$	$y^{(i)}$
$\vdots$	$\vdots$	$\vdots$
N	$\cdots$	.

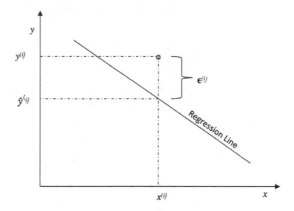

FIGURE 2.2: Error at point $x^{(i)}$

errors. Thus, when we fit a regression line, each point may actually be not sitting on the line, but away from it. Thus, the fit of each point to the regression line may have an error associated with it. Let us call the error or residual associated with the fit of the ith point $\epsilon^{(i)}$. This is illustrated in Figure 2.2. An assumption that we make about the spread of the residuals is that they are normally distributed, or they obey the Gaussian distribution. Since our dataset has n points in it, the total cumulative error of fit of a regression line to a training dataset can be written as

$$E = \sum_{i=1}^{N} \epsilon^{(i)} \tag{2.1}$$

if we simply add all the errors across the dataset. Thus, we can write an objective function to obtain the regression line as

Find line $y = \theta_1 x + \theta_0$ such that it minimizes the cumulative error $E = \sum_{i=1}^{N} \epsilon^{(i)}$.

However, the problem with this approach is that some of the errors are likely to be positive and some of the errors negative, and negative errors will cancel positive errors, and the regression line that minimizes the direct sum of errors may actually turn out to be a bad fit. An alternative may be to compute the cumulative error as

$$E = \sum_{i=1}^{N} |\epsilon^{(i)}| \tag{2.2}$$

where $|.|$ is the absolute value. This representation of cumulative error is good and often used, but absolute values are usually difficult to deal with in mathematics. Thus, another alternative approach that has been used and that we use as the cumulative error is the sum of the squares of individual errors, given by

$$E = \sum_{i=1}^{N} \left\{ \epsilon^{(i)} \right\}^2. \tag{2.3}$$

Now, since squares of both positive and negative errors are positive, we have a total error that does not vanish unless there is no cumulative error at all. If there is error, depending on the magnitude of the individual errors (less than 1 or more than 1), the error can be squashed or magnified. This is because numbers with absolute values bigger than 1 become even bigger if squared, and numbers with absolute values less than 1 become still smaller when squared. Normalization of values may help mitigate this issue, but we do not discuss it here. The modified objective function to obtain the regression line becomes

Find line $y = \theta_1 x + \theta$ such that it minimizes the cumulative error
$E = \sum_{i=1}^{N} \left\{ \epsilon^{(i)} \right\}^2.$

Because there is potentially an error or residual $\epsilon^{(i)}$ at each point, we can write

$$y^{(i)} = \theta_1 x^{(i)} + \theta_0 + \epsilon^{(i)}. \tag{2.4}$$

Since the error $\epsilon^{(i)} = y^{(i)} - \theta_1 x^{(i)} - \theta_0$, we can write the cumulative error expression as

$$E = \sum_{i=1}^{N} \left\{ y^{(i)} - \theta_1 x^{(i)} - \theta_0 \right\}^2.$$

This error E is also called loss in machine learning, although this term is not so commonly used in mathematics. In this expression for the cumulative error, the $x^{(i)}$ and $y^{(i)}$ values are known from the training dataset. The values for θ_1 and θ_0 that minimize E need to be obtained to find a line that fits the data well. Thus, to find the equation of the line that minimizes this cumulative squared error, we need to obtain its partial derivatives with respect to the two "variables" θ_1 and θ_0, set them to 0 and solve for the values of θ_1 and θ_0.

$$0 = \frac{\partial E}{\partial \theta_1}$$

$$= \frac{\partial}{\partial \theta_1} \sum_{i=1}^{N} \left\{ y^{(i)} - \theta_1 x^{(i)} - \theta_0 \right\}^2$$

$$= 2 \sum_{i=1}^{N} \left\{ y^{(i)} - \theta_1 x^{(i)} - \theta_0 \right\} \frac{\partial}{\partial \theta_1} \sum_{i=1}^{N} \left\{ y^{(i)} - \theta_1 x^{(i)} - \theta_0 \right\}$$

$$= -2 \sum_{i=1}^{N} \left\{ y^{(i)} - \theta_1 x^{(i)} - \theta_0 \right\} x^{(i)}$$

$$= \sum_{i=1}^{N} \left\{ y^{(i)} - \theta_1 x^{(i)} - \theta_0 \right\} x^{(i)}$$

This gives us a linear equation in θ_1 and θ_0 as given below:

$$\sum_{i=1}^{N} y^{(i)} x^{(i)} - \left(\sum_{i=1}^{N} \left\{ x^{(i)} \right\}^2 \right) \theta_1 - \left(\sum_{i=1}^{N} \left\{ x^{(i)} \right\} \right) \theta_0 = 0 \qquad (2.5)$$

Similarly, we have to set the partial derivative of E w.r.t. θ_0 to 0 as well.

$$0 = \frac{\partial E}{\partial \theta_0}$$

$$= \frac{\partial}{\partial \theta_0} \sum_{i=1}^{N} \left\{ y^{(i)} - \theta_1 x^{(i)} - \theta_0 \right\}^2$$

$$= 2 \sum_{i=1}^{N} \left\{ y^{(i)} - \theta_1 x^{(i)} - \theta_0 \right\} \frac{\partial}{\partial \theta_0} \sum_{i=1}^{N} \left\{ y^{(i)} - \theta_1 x^{(i)} - \theta_0 \right\}$$

$$= -2 \sum_{i=1}^{N} \left\{ y^{(i)} - \theta_1 x^{(i)} - \theta_0 \right\}$$

$$= \sum_{i=1}^{N} \left\{ y^{(i)} - \theta_1 x^{(i)} - \theta_0 \right\}$$

This can be rewritten as another linear equation in θ_1 and θ_0.

$$\sum_{i=1}^{N} y^{(i)} - \left(\sum_{i=1}^{N} \left\{ x^{(i)} \right\} \right) \theta_1 - \sum_{i=1}^{N} \theta_0 = 0$$

$$\sum_{i=1}^{N} y^{(i)} - \left(\sum_{i=1}^{N} \left\{ x^{(i)} \right\} \right) \theta_1 - N \theta_0 = 0 \qquad (2.6)$$

Given two linear equations (Equations 2.5 and 2.6) in θ_1 and θ_0, we can solve for them and obtain the equation of the line.

The solutions we get are

$$\theta_1 = \frac{N\left(\sum_{i=1}^{N}\left\{x^{(i)}y^{(i)}\right\}\right) - \left(\sum_{i=1}^{N}\left\{x^{(i)}\right\}\right)\left(\sum_{i=1}^{N}y^{(i)}\right)}{N\left(\sum_{i=1}^{N}\left\{x^{(i)}\right\}^2\right) - \left(\sum_{i=1}^{N}\left\{x^{(i)}\right\}\right)^2} \tag{2.7}$$

$$\theta_0 = \frac{1}{N}\sum_{i=1}^{N}y^{(i)} - \theta_1\frac{1}{N}\sum_{i=1}^{N}x^{(i)}$$

$$= \overline{y} - \theta_1\overline{x}. \tag{2.8}$$

In Equation 2.8, $\overline{y}$ and $\overline{x}$ are means. The computation is described in terms of an algorithm in Algorithm 2.1.

Algorithm 2.1: Algorithm for Linear Least Squares Regression with 1 independent variable

---

Input: Training Set T[1..2] with two columns
1   X ←T[1] //X is an array containing the entire first column of T
2   N ← length(X)
3   Y ← T[2] //Y is an array containing the entire second column of T
4   XSum ← 0
5   YSum ← 0
6   XYSum ← 0
7   XSquaredSum ← 0
8   //Loop over the entire dataset
9   **for** *i = 1 to N do* **do**
10   XSum ← XSum + X[i]
11   XSquaredSum ← XSquaredSum + X[i] * X[i]
12   YSum← YSum + Y[i]
13   XYSum ← XYSum + X[i]*Y[i]
14   theta1← (N * XYSum - XSum * XSum) / (N * XSquaredSum - XSum * XSum)
15   theta0 ← (Ysum - theta1 * Xsum)/N
16   **return** *theta0, theta1*

Although we know the algorithm for linear least squares regression, we generally do not program it from scratch. We can perform linear least squares regression using a statistical package like R. We obtained a scatter plot of the Drugs and Grades data first. The scatter plot can be seen in Figure 2.1. We performed a linear regression fit that can be seen in Figure 2.1. We followed it by a summary description of the least squares fit. The output from R is given in Figure 2.1. The equation of the line given by R is

$$\hat{y} = -0.09743x + 9.21308. \tag{2.9}$$

So, this line can be used as a trend line to explain the data in the training dataset. This line can also be used to predict the estimated value of $\hat{y}$ of y, given an arbitrary value of x. As noted earlier, most actual points in the dataset do not lie on this line. Thus, at each point, we are likely to have a non-zero error or residual. However, if we assume that the underlying relationship is linear, this is the line of best fit, considering the summed squares of errors. In other words, among all possible lines we can draw to describe the trend in the data, this line has the smallest sum of squares of errors. The slope of this line, also called the least squares regression line (LSRL), is negative, meaning that the score in math rises as the LSD concentration goes down. The intercept of 9.21 is the highest score that a student can receive if the LSD concentration is 0. Given LSD values of 35 and 75, we can compute the predicted math scores to be 5.80 and 1.91, respectively, using Equation 2.9.

2.2   Evaluating Regression

It is not a good idea to create a learned model, which is the LSRL regression line here, and accept it without any quantified measurement of goodness. There are various metrics that are used to evaluate how well a trained regression model fits the data. A few of them are discussed below.

2.2.1   Coefficient of Determination R^2

A common way to determine how well a regression model fits the data is by computing a statistical measure called the coefficient of determination, also called R^2. We compute R^2 with respect to a so-called baseline model that represents the "worst" regression model possible. This baseline model does not use the independent variables (in mathematical terms, features in machine learning terms) to predict the dependent variable (here, the label). It simply uses the mean of the dependent variable (here, label y) and always predicts it as the value of $\hat{y}$. A regression model that we fit such as the LSRL, described here, is compared to the baseline model to quantify the goodness of fit. In other words, R^2 explains how good our model (here, LSRL) is when compared to the baseline model. In Figure 2.3, the horizontal line is the baseline model that always predicts $\bar{y}$ as the value of $\hat{y}$, given any value of x, irrespective of what it is. The LSRL line is the fitted model which uses the independent variable (here, feature x) to predict the value of the dependent variable (here, the label y).

Let SSE be the sum of squared errors of our regression model. Each error is the error in prediction for the corresponding y value, given an $x^{(i)}$ value. $y^{(i)}$ is the actual value of y at $x^{(i)}$ in our training dataset and $\hat{y}^{(i)}$ is the predicted

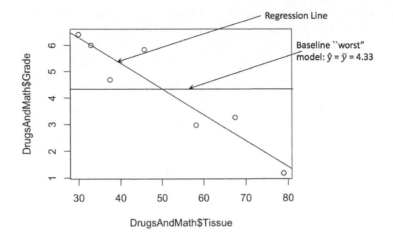

FIGURE 2.3: Regression line and Baseline model

value of y at $x^{(i)}$.

$$SSE = \sum_{i=1}^{N} \left(y^{(i)} - \hat{y}^{(i)} \right)^2. \tag{2.10}$$

Let SST be the sum of squared errors for the baseline model, called the Squared Sum for Total error (or, total squared sum error):

$$SST = \sum_{i=1}^{N} \left(y^{(i)} - \overline{y} \right)^2. \tag{2.11}$$

This is called total error because we assume that the baseline model is the worst model and we cannot create another model that is worse. The statistic R^2 is given as

$$R^2 = 1 - \frac{SSE}{SST} \tag{2.12}$$

$$= 1 - \frac{\sum_{i=1}^{N} \left(y^{(i)} - \hat{y}^{(i)} \right)^2}{\sum_{i=1}^{N} \left(y^{(i)} - \overline{y} \right)^2}. \tag{2.13}$$

Here, $\frac{SSE}{SST}$ is a ratio that tells us what fraction of the worst total error our regression model makes. Therefore, $1 - \frac{SSE}{SST}$ is the fraction that tells us how much better our model is compared to the worst model.

The worst possible regression model is the baseline model itself. In such a case,

$$R^2 = 1 - 1 = 0. \tag{2.14}$$

The "best" regression model would be the one that fits all data points perfectly, i.e., all points fall on the regression line with individual errors of 0 each.

Our actual LSRL model (Figure 2.3), in this case, does not look very close to such a "best" model. But, if we were able to find such a "best" model, SSE would be 0 for it. As a result,

$$R^2 = 1 - 0 = 1 \tag{2.15}$$

for this model. Thus, R^2 ranges between 0 and 1, with such an assumption regarding the baseline or the "worst" model. If R^2 is close to 1, the regression model fits the data well, and if it is close to 0, it does not fit the data well. Another way of saying it is that $R^2 \approx 1$ explains the variations in the dependent variable (here, the label) y well, and $R^2 \approx 0$ does not explain the variation in y well at all. If for a regression model, the value of R^2 is r^2, we say that the regression model explains $r^2\%$ of the variations in y, given values of x. Thus, when the R system says the R-squared value is 0.8778 for the Drugs and Grades dataset, it means that 87.78% of the variation in the math grade is captured by the model.

2.2.2 Adjusted R^2

The R^2 metric is simple and has many associated problems as a metric for goodness of fit of a regression model. One complaint about R^2 is that it has been observed as the number of independent variables (also called explanatory variables in statistics, we call them features in machine learning) increases, the value of R^2 increases, possibly weakly. To take this into account, a metric called Adjusted R^2 or R^2_{adj} is also often computed. To motivate R^2_{adj}, we revisit the formula for R^2 and write it out a bit differently as follows:

$$R^2 = 1 - \frac{SSE}{SST} \tag{2.16}$$

$$= 1 - \frac{\frac{1}{N}SSE}{\frac{1}{N}SST} \tag{2.17}$$

$$= 1 - \frac{VAR_{SSE}}{VAR_{SST}}. \tag{2.18}$$

Here, $\frac{1}{N}SSE = \frac{1}{N}\sum_{i=1}^{n}\left(y^{(i)} - \hat{y}^{(i)}\right)^2$ is the standard definition of variance of the residuals or errors, which we can call VAR_{SSE}, also written as s^2_{SSE}. We consider all N residuals in computing VAR_{SSE}. Also, $\frac{1}{N}SST = \frac{1}{N}\sum_{i=1}^{n}\left(y^{(i)} - \overline{y}\right)^2 = VAR_{SST}$ where VAR_{SST} (also written as s^2_{SST}) is the variance of the total error terms or errors from the baseline or horizontal mean y model. There are two variances in the formula for R^2 above. To write an improved version of R^2, which we call R^2_{adj}, we bring in the degrees of freedom of these variances, instead of simply writing N in the denominator in the variance definitions.

$$R^2_{adj} = 1 - \frac{\frac{VAR_{SSE}}{df_e}}{\frac{VAR_{SST}}{df_t}} \tag{2.19}$$

where $df_e = N - 1$ is the number of degrees of freedom when computing variance of N things, and $df_t = N - q - 1$ is the number of degrees of freedom when computing the error variance with respect to a regression model (here, linear), where q is the number of explanatory (or independent) variables, which is actually the number of features in our parlance. The degrees of freedom for the error terms with respect to a regression model must take into account the number of parameters used by the model. For the linear regression model in this case, there is one independent variable x. Thus, $q = 1$. As a result, the formula for R_{adj}^2 for LSRL is:

$$R_{adj}^2 = 1 - \frac{\frac{VAR_{SSE}}{N-1}}{\frac{VAR_{SST}}{N-2}}. \qquad (2.20)$$

Obviously, if N is large, having N in the denominator or $N - 1$ or $N - 2$ does not make much difference. But, if N is small, there may be a visible difference between the values of R^2 and R_{adj}^2.

In our example, R computes R_{adj}^2 as 0.8534, which is smaller than R^2 whose value is 0.8778.

2.2.3   F-Statistic

The F-Statistic is another value that is commonly computed to gauge the quality of a regression fit. We will first discuss how F-statistic is computed, and then how it is used. Earlier, we defined the Sum of Squared Errors, SSE as

$$SSE = \sum_{i=1}^{N} \left(y^{(i)} - \hat{y}^{(i)} \right)^2. \qquad (2.21)$$

We can compute the Mean Squared Error, MSE as

$$MSE = \frac{SSE}{N - p} \qquad (2.22)$$

where p is the number of parameters, which are θ_1 and θ_0. Thus, $p = q + 1$, where p is the number of independent variables or features. $p = 2$ for linear regression in our example. Then, for linear regression,

$$MSE = \frac{SSE}{N - 2}, \qquad (2.23)$$

and we say that it has a degree of freedom of $N - 2$. This definition reflects the fact that to begin with we have N variables or degrees of freedom in the description of the data. Since there are two coefficients, two degrees of freedom are covered by the linear regression line. We can define a quantity called Sum of Squares Regression, SSR, which is the sum of squares of errors for each predicted y value from the baseline "worst" model, as follows:

$$SSR = \sum_{i=1}^{n} \left(\hat{y}^{(i)} - \overline{y} \right)^2. \qquad (2.24)$$

It can be shown that

$$SST = SSR + SSE. \tag{2.25}$$

Now, let

$$MSR = \frac{SSR}{q-1}, \tag{2.26}$$

with a degree of freedom $p = q - 1$. The concept of degrees of freedom takes some effort to understand clearly, but statisticians depend on them for the formulas to work correctly. F-statistic is defined as

$$F\text{-}statistic = \frac{MSR}{MSE}. \tag{2.27}$$

Given the degrees of freedom for the two quantities MSR and MSE, both of which are variances, we can use these values to draw a F-distribution graph. The F-distribution is a distribution of the ratio of two variances, and it is a graph which is single-tailed in the right side, or upper-tailed. We can draw the graph using R packages such as `visualize` or `vistributions` that allow one to draw probability distributions. The F-distribution needs two parameters, the degrees of freedom of the variance in the numerator and the variance in the denominator. The graph rises quickly from 0, and has the highest value near 0, and then it slowly falls. This graph has F-value on the X-axis and the probability on the Y-axis. The F-statistic, which we compute for our dataset, is a particular F-value on the X-axis of this graph that indicates the point at which the F-value ratio between the two variances for this specific training dataset falls. The value of F-statistic computed by R for our example is 35.39, with degree of freedom 1 (one independent variable x) for numerator and degree of freedom 5 (for 7 data points - 2) for the denominator. A graph that shows what a plot of F-distribution looks like is given in Figure 2.4(a).This is not the real graph that comes from our computation in this case, but it shows the important components of the graph more clearly. An actual plot is given in Figure 2.4(b).

2.2.3.1   Is the Model Statistically Significant? F-Test

It is customary to compute what is called a p-value when we perform a statistical test to indicate if the results obtained are statistically significant, i.e., they have any (statistical) merit. That is, whether the results can be trusted and are meaningful, and can really be used. The p-values are computed in different ways for different situations.

In addition, when we perform a statistical test, there is always an associated null hypothesis and an alternate hypothesis. For linear regression, the null hypothesis is that the coefficients associated with the independent variables (or, features) is 0, i.e., the regression coefficients, except the constant or intercept, are all 0. In other words, the null hypothesis says there is no relationship of any significance between the dependent and independent variables. The alternate hypothesis is that the coefficients are not equal to 0, i.e., there

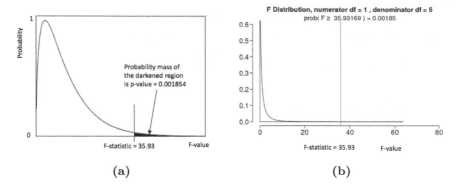

FIGURE 2.4: F-statistic and p-value, with F-statistic $= 35.39$ for an F-distribution with degrees of freedom 1 and 5. The p-value obtained is 0.001854, which is less than the assumed value of α, making the regression fit statistically significant. (a) For illustration, (b) Actual drawn at https://shiny.rit.albany.edu/stat/fdist/.

is actually a linear relationship between the dependent variable and the independent variable(s). In particular, for linear regression with one independent variable, the null hypothesis would be that $\theta_1 = 0$. The alternate hypothesis is that $\theta_1 \neq 0$. We have to be able to reject the null hypothesis and accept the alternate hypothesis to accept the results obtained for linear regression by a system like R. We can reject the null hypothesis by performing what is called the F-test.

The F-test compares our model (LSRL in this case) with the intercept only model and decides if the added coefficients (here, just θ_1) improved the model. To perform the F-test, we compute the F-statistic, discussed earlier, for our dataset. We also compute what is called a p-value, which is the cumulative probability with an F-value (the ratio) less than or equal to the F-statistic we compute for our dataset is likely. This p-value is computed using the F-distribution that plots the probabilities of the ratio of two variances we discussed earlier, when discussing the F-statistic. In addition, to perform the F-test, we assume what is called an α, which is a significance level. α is simply a small number (e.g., $\alpha = 0.05$). If the p-value computed for the F-statistic comes out to be less than or equal to α, the F-test says that we can reject the null hypothesis and accept the alternate hypothesis. If the null hypothesis cannot be rejected, we should not consider this test any further and say that the regression model is not significant in explaining the data. The p-value corresponding to the F-statistic we obtain for our dataset is the probability mass in the F-distribution to the right of the F-statistic or the area under the probability curve to the right of the F-statistic $= 35.39$. This is illustrated in Figure 2.4. If the null hypothesis can be rejected, the linear regression model we computed is meaningful in explaining the fit of the data, and can be used for predictive purposes. In this case, the p-value comes out to 0.001854 or 0.1854%, as seen in Figure 2.4. If we take that $\alpha = 0.05$, which is a value normally assumed, the p-value is less than α and as a result, we can reject the null hypothesis. In other words, we can consider that the regression line $\hat{y} = -0.09743473x + 9.21308462$ obtained by R for our dataset is a good fit. When performing regression, we do not have to explicitly say that we reject the null hypothesis (assuming p-value $\leq \alpha$) every time, but we should know what transpires in the background so that we accept the regression line computed. The F-statistic is not really used except to compute the p-value from the F-distribution curve.

2.2.4 Running the Linear Regression in R

Below, we show the R code that performs LSRL fit for the LSDMath dataset.

```
1 LSDMath <- read.csv("~/Datasets/txt/LSDMath.txt")
2 View(LSDMath)
3 plot(LSDMath)
4 myfit <- lm (Math ~ LSD, data=LSDMath)
5 abline(myfit)
6 summary(myfit)
```

First, we load the LSDMath dataset from a specific location on our computer. It is a comma-separated value (CSV) file. The dataset is read and loaded in to a variable called `LSDMath`. We can view the dataset on the terminal as a table using the `View` command. We scatterplot it using the `plot` command. To perform the LSLR regression, we use the `lm` command and save details of the regression line in the variable `myfit`. `abline` draws the LSRL line seen in Figure 2.3, and `summary` prints the details of the LSRL fit. Below, we show the summary of the fit as printed by R.

```
Residuals:
         1          2          3          4          5          6          7
-0.352561  -0.572383   0.620837  -0.872205   1.064811  -0.005533   0.117034

Coefficients:
            Estimate Std. Error t value Pr(>|t|)
(Intercept)  9.21308    0.86101  10.700 0.000123 ***
LSD         -0.09743    0.01626  -5.994 0.001854 **
---
Signif. codes:  0 '***' 0.001 '**' 0.01 '*' 0.05 '.' 0.1 ' ' 1

Residual standard error: 0.741 on 5 degrees of freedom
Multiple R-squared:  0.8778,      Adjusted R-squared:  0.8534
F-statistic: 35.93 on 1 and 5 DF,  p-value: 0.001854
```

There are seven points in our dataset. When LSRL is fitted, each point has an error or residual. They are printed out by R. The coefficient for x, which happens to be the LSD variable, is -0.09743. The intercept is 9.21308. This gives us the equation for the fitted line to be: $y = -0.09743x + 9.21308$. The RSS value is 0.741 with 5 degrees of freedom (7 points - 2). The computed R^2 is 0.8778 and R^2_{adj} is 0.8534. The F-statistic is 35.93 with a numerator and denominator degrees of freedom 1 and 5, respectively. The p-value for the fit of the LSRL is 0.001854, which makes the regression line an acceptable fit to the dataset.

In addition to the "estimates" of the values of the parameters (intercept and the coefficient for x or LSD), R also gives a few additional values for the estimates in a small table. When the actual algorithm finds these values for the parameters, it goes through a process that looks at various alternative values and chooses one estimate for a parameter, considering how reliable or significant the value is. It uses what is called the t-test for evaluating significance in this case. However, we do not overwhelm the reader with details of another statistical test of significance. The main thing to do is to look at the column labeled Pr (>|t|), which says Intercept has a p-value of 0.000123, with a significance code of ***, making the estimate quite reliable because p-value ≤ 0.05. The estimate of the coefficient for LSD is also good with a p-value of 0.001554 and significance code of **. These p-values are computed differently compared to the p-value for F-statistic and we will discuss this later in the book. The note to be made is that we have a good estimate for each of the parameter values, and we also have a good overall LSRL model.

2.2.5 The Role of Optimization in Regression

An astute reader may have noticed that we have to solve an optimization problem to obtain the least squared error line of fit, given a dataset with an independent variable and an associated dependent variable. This is not unsual at all in machine learning. A machine learning algorithm attempts to obtain a good model to that fits the data, keeping in mind that the purpose is to build a model that learns the essence of the particulars of the training dataset, but also focuses on generalizing to examples not in the training dataset. This process usually or always involves an underlying optimization algorithm. In fitting an LSRL, we used a simple optimization algorithm that we learn in calculus. It involves obtaining partial derivatives of an expression or function with respect to its variables, setting the derivatives to 0, obtaining a system of linear equations and solving them. In the case where we have a single independent variable, this approach results in a quick solution to the optimization problem. However, we will see that the underlying optimization problem becomes more difficult as our machine learning algorithm becomes more complex. For example, it is possible that the number of independent variables becomes large, say in the hundreds or thousands. There is always one dependent variable. In such a situation, we will have to solve a system of linear equations in a large number of variables. Solving such a system and obtaining reliable and stable results is not straightfoward. This is where, we may start approaching the optimization problem for fitting a good line or plane or hyperplane or some other curve to a training dataset using different and more sophisticated algorithms. There are many ways to solving them. This is an active area of research in mathematics. A variety of approaches such as gradient descent, QR factorization, stochastic gradient descent, semi-definite optimization, and conic optimization can be used. We will discuss one or more of these algorithms later in the book as necessary. In particular, we should note that the `lm` implementation in R does not use the approach of partial derivatives and systems of linear equations discussed here. It uses a more sophisticated approach called QR matrix decomposition to solve an over-determined system of equations. We discussed the approach we choose here to introduce the idea that optimization plays a significant and all pervasive role in machine learning, and a simple algorithm that everyone understands from calculus is a good place to start. We will discuss a few other optimization approaches later in this book, as needed.

2.3 Multi-Dimensional Linear Regression

So far, we have discussed regression where we have a single independent variable, and a single dependent variable. However, it is quite likely that we have several independent variables or features to describe each example in a training dataset. Thus, the ith example is given as

$$\mathbf{x}^{(i)} = [x_1^{(i)}, \cdots x_j^{(i)} \cdots x_n^{(i)}]$$

where $x_i^{(i)}$ is the jth feature's value for the ith example.

In such a case, when we fit a linear regression model, the equation of the model is

$$y = \theta_0 + \theta_1 x_1 + \cdots + \theta_j x_j + \cdots + \theta_n x_n \tag{2.28}$$

where

$$\vec{\theta} = [\theta_0, \theta_1, \cdots, \theta_j, \cdots, \theta_n]$$

is the vector of coefficients in the fitted line. In other words, they are the parameters that need to be found to do the fitting.

When we look at the ith point $[x_1^{(i)}, \cdots, x_n^{(i)}]$, its y value $y^{(i)}$ is given as follows:

$$y^{(i)} = \theta_0 + \theta_1 x_1(i) + \cdots + \theta_j x_j(i) + \cdots + \theta_n x_n(i) + \epsilon^{(i)}$$

where $\epsilon^{(i)}$ is the error in prediction by the fitted line, plane or hyperplane. In other words, the error in fit of the ith training data point is given as

$$\epsilon^{(i)} = y^{(i)} - \theta_0 - \theta_1 x_1^{(i)} - \cdots - \theta_j x_j^{(i)} - \cdots - \theta_n x_n^{(i)}. \tag{2.29}$$

Just like fitting a regression line when we had scalar x, we find the equation of the plane in n dimensions, for which the sum of the squares of the individual errors for each training data point is minimized.

Thus, the objective function to obtain the regression plane or hyperplane becomes

Find hyperplane $y = \theta_0 + \theta_1 x_1 + \cdots + \theta_j x_j + \cdots + \theta_n x_n$ such that it minimizes the cumulative error $E = \sum_{i=1}^{n} \left\{ \epsilon^{(i)} \right\}^2$.

Like before, we have to compute $\frac{\partial E}{\partial \theta_0} \cdots \frac{\partial E}{\partial \theta_j} \cdots \frac{\partial E}{\partial \theta_n}$ and set each one to 0.

Let us just compute $\frac{\partial E}{\partial \theta_0}$ here and leave the others as exercise.

$$
\begin{aligned}
0 &= \frac{\partial E}{\partial \theta_0} \\
&= \frac{\partial}{\partial \theta_0} \left\{ \epsilon^{(i)} \right\}^2 \\
&= \frac{\partial}{\partial \theta_0} \left\{ y^{(i)} - \theta_0 - \theta_1 x_1^{(i)} - \cdots - \theta_j x_j^{(i)} - \cdots - \theta_n x_n^{(i)} \right\}^2 \\
&= 2 \left\{ y^{(i)} - \theta_0 - \theta_1 x_1^{(i)} - \cdots - \theta_j x_j^{(i)} - \cdots - \theta_n x_n^{(i)} \right\} \\
&\qquad \frac{\partial}{\partial \theta_0} \left\{ y^{(i)} - \theta_0 - \theta_1 x_1^{(i)} - \cdots - \theta_j x_j^{(i)} - \cdots - \theta_n x_n^{(i)} \right\} \\
&= -2 \left\{ y^{(i)} - \theta_0 - \theta_1 x_1^{(i)} - \cdots - \theta_j x_j^{(i)} - \cdots - \theta_n x_n^{(i)} \right\} \\
&= \left\{ y^{(i)} - \theta_0 - \theta_1 x_1^{(i)} - \cdots - \theta_j x_j^{(i)} - \cdots - \theta_n x_n^{(i)} \right\} \qquad (2.30)
\end{aligned}
$$

By partially differentiating E with respect to $n+1$ variables, we get $n+1$ equations in $n+1$ unknowns, $\theta_0 \cdots \theta_n$. We can solve them using any method for solving a system of linear equations and obtain the equation of the hyperplane that is the regression model that fits the training dataset the best.

2.3.1 Multi-dimensional Linear Least Squares Regression Using R

Running multi-dimensional regression in R is as simple as running two-dimensional regression. Below, we show the code that is used to run multi-dimensional LSRL with the immigrant dataset we discussed earlier.

```
1 Immigrants <- read.csv("~/Datasets/txt/Immigrants.csv")
2 View(Immigrants)
3 plot(Immigrants)
4 myfit <- lm (Income ~ English + Literacy + Residency, data=Immigrants)
5 summary(myfit)
```

The dataset is stored in a file called `Immigrants.txt`. It is read and stored in the variable called `Immigrants`. Its contents can be seen as a table in R

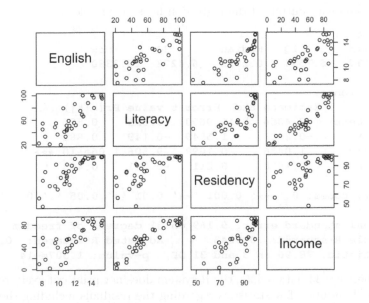

FIGURE 2.5: Scatterplot of pair-wise relationships among features in the Immigrants dataset

(the examples were produced using R Studio) using the `View` command. This dataset has 3 features named `English`, `Literacy` and `Residency`. The plot of the dataset by R is shown in Figure 2.5. When we have multiple variables to plot, the `plot` command in R plots all combinations of two variables at a time. The set of plots is organized in a matrix and each plot is repeated twice. The first row shows how `English` is scatterplotted against `Literacy`, `Residency` and `Income`. The second column scatterplots `Literacy` against the other three variables, and so on. Such plots let us view if there are clear linear relationships or collinearity between any two variables, and determination of such relationships may help us do a better job of performing regression, although we do not discuss the issue much here. In this case, the dependent variable `Income` is approximately linearly correlated to the dependent variable `Literacy`, and hence, we expect to see a positive coefficient for it in the LSRL equation. Line 4 of the code performs multi-dimensional LSRL regression of the `Income` dependent variable in terms of the three features or independent variables. It fits a linear equation of the form $\hat{y} = \theta_0 + \theta_1 x_1 + \theta_2 x_2 + \theta_3 x_3$ where $x_1 = $ `English`, $x_2 = $ `Literacy`, $x_3 = $ `Residency`, and $y = $ `Income`. There are three features (independent variables): x_1, x_2 and x_3. The dependent variable is y. $\hat{y}$, which is predicted y, is a plane in three dimensions. Such planes in more than three dimensions are usually called hyperplanes.

The summary output of the regression fit is given below:

```
Residuals:
     Min        1Q    Median        3Q       Max
-14.6517   -6.7534    0.2922    6.4278   16.3390

Coefficients:
              Estimate Std. Error t value Pr(>|t|)
(Intercept)     7.4864    11.9690   0.625     0.536
English        -0.2318     1.5549  -0.149     0.882
Literacy        1.0436     0.1098   9.506  1.06e-10 ***
Residency      -0.1568     0.2115  -0.741     0.464
---
Signif. codes:  0 '***' 0.001 '**' 0.01 '*' 0.05 '.' 0.1 ' ' 1

Residual standard error: 9.169 on 31 degrees of freedom
Multiple R-squared:  0.8843,       Adjusted R-squared:  0.8731
F-statistic: 78.96 on 3 and 31 DF,  p-value: 1.307e-14
```

Since there are 33 data points, the R system does not print out every residual or error. It prints a few statistics regarding the residuals including the minimum, maximum and median values. It also prints the first quartile and third quartile values of the residuals. The coefficients of the linear fit for the three features English, Literacy and Residency are printed with some additional details. These details tell us that R is able to estimate the coefficient for only Literacy reliably, with a p-value of 1.06×10^{10}. The other estimated coefficients have p-values of 0.464, 0.536 and 0.882, and hence, much above the usual cutoff of 0.05. The R^2 value is 0.8843, the R^2_{adj} value is 0.8731, the F-statistic is 78.96 with 3 parameters (for θ_1, θ_2 and θ_3) and 31 (33 data points - 2) as degrees of freedom. The linear regression fit is given as

$$\hat{y} = -0.2318x_1 + 1.0436x_2 - 0.1568x_3 + 7.4864 \qquad (2.31)$$

with a p-value of $1.307\ e^{-14}$, a very small value, making the overall fit line statistically acceptable, although three of the coefficients are not stable or reliable.

2.4 Polynomial Regression

Instead of fitting a linear function, we can also fit a polynomial of higher degree to our data. Once again, like our simple example in the beginning of the chapter, we assume we have only one independent variable. The independent variable is x and the dependent variable is y. In some cases, just by looking at the data, we may be able to surmise that a linear relationship does not exist between the dependent and independent variables. In such a situation, we can attempt to fit polynomials of several degrees to see what fits the data the best.

In general, we can fit an nth degree polynomial:

$$\hat{y} = \theta_0 + \theta_1 x + \cdots + \theta_n x^n \tag{2.32}$$

to a dataset. As an illustrative example, let us try to fit a second degree polynomial

$$\hat{y} = \theta_0 + \theta_1 x + \theta_2 x^2. \tag{2.33}$$

Since there is an error of fit at every point, we can write for the ith point,

$$y^{(i)} = \theta_0 + \theta_1 x^{(i)} + \theta_2 \left(x^{(i)}\right)^2 + \epsilon^{(i)}.$$

Then, the error of fit at the ith training point is

$$\epsilon^{(i)} = y^{(i)} - \theta_0 - \theta_1 x^{(i)} - \theta_2 \left(x^{(i)}\right)^2.$$

Here $x^{(i)}$ is the x value for the ith data example, $\epsilon^{(i)}$ is the error or residual for this example, and $y^{(i)}$ is the y value at $x^{(i)}$ in the actual data example. $\hat{y}^{(i)}$ is the value predicted for y at x by the second degree regression polynomial. The general optimization problem to solve becomes:

Find $y = \theta_0 + \theta_1 x + \cdots + \theta_n x^n$ *such that it minimizes the cumulative error* $E = \sum_{i=1}^{n} \left\{\epsilon^{(i)}\right\}^2$,

and the specific optimization problem for the second degree case becomes

Find $y = \theta_0 + \theta_1 x + \theta_2 x^2$ *such that it minimizes the cumulative error* $E = \sum_{i=1}^{n} \left\{\epsilon^{(i)}\right\}^2$,

Like before, we have to compute $\frac{\partial E}{\partial \theta_0}$, $\frac{\partial E}{\partial \theta_1}$ and $\frac{\partial E}{\partial \theta_2}$ and set each one to 0. Let us just compute one of these partial derivatives here: $\frac{\partial E}{\partial \theta_2}$.

$$
\begin{aligned}
0 &= \frac{\partial E}{\partial \theta_2} \\
&= \frac{\partial}{\partial \theta_2} \left\{\epsilon^{(i)}\right\}^2 \\
&= \frac{\partial}{\partial \theta_2} \left\{y^{(i)} - \theta_0 - \theta_1 x^{(i)} - \theta_2 \left(x^{(i)}\right)^2\right\}^2 \\
&= 2\left\{y^{(i)} - \theta_0 - \theta_1 x^{(i)} - \theta_2 \left(x^{(i)}\right)^2\right\} \\
&\qquad \frac{\partial}{\partial \theta_2} \left\{y^{(i)} - \theta_0 - \theta_1 x^{(i)} - \theta_2 \left(x^{(i)}\right)^2\right\} \\
&= -2\left\{y^{(i)} - \theta_0 - \theta_1 x^{(i)} - \theta_2 \left(x^{(i)}\right)^2\right\} \left(x^{(i)}\right)^2 \\
&= \left\{y^{(i)} - \theta_0 - \theta_1 x^{(i)} - \theta_2 \left(x^{(i)}\right)^2\right\} \left(x^{(i)}\right)^2
\end{aligned} \tag{2.34}
$$

Similarly, we can obtain two other equations by setting the other two partial derivatives to 0. We will get the following.

$$0 = \frac{\partial E}{\partial \theta_0} = y^{(i)} - \theta_0 - \theta_1 x^{(i)} - \theta_2 \left(x^{(i)}\right)^2$$

$$0 = \frac{\partial E}{\partial \theta_1} = \left\{y^{(i)} - \theta_0 - \theta_1 x^{(i)} - \theta_2 \left(x^{(i)}\right)^2\right\} x^{(i)}$$

Thus, we have three linear equations in three unknowns, θ_0, θ_1 and θ_2. We can solve to obtain the values of θ_0, θ_1 and θ_2 and obtain the second degree function we want to fit to the training dataset.

In general, if we have an nth degree polynomial, we get

$$\epsilon^{(i)} = y^{(i)} - \sum_{j=0}^{n} \theta_j \, x^j.$$

Therefore,

$$E = \sum_{i=1}^{N} \left\{\epsilon^{(i)}\right\}^2$$

$$= \sum_{i=1}^{N} \left(y^{(i)} - \sum_{j=0}^{n} \theta_j x^j\right),$$

and

$$0 = \sum_{i=1}^{N} \left(y^{(i)} - \sum_{j=0}^{n} \theta_j x^j\right) \left(x^{(i)}\right)^j$$

for $j = 0 \cdots n$. We have $n + 1$ equations in $n + 1$ unknowns, and we can solve for $\theta_0 \cdots \theta_n$ to obtain the polynomial regression.

Another approach, which is equivalent, is the following. Let us look at the quadratic equation that we want to fit: $y = \theta_0 + \theta_1 x + \theta_2 x^2$. Also define two variables $z_1 = x$ and $z_2 = x^2$. Then the equation to fit becomes

$$y = \theta_0 + \theta_1 z_1 + \theta_2 z_2.$$

We can simply solve for the least squares regression line like we did for the multiple regression case. The only thing that happened is that the variable names have changed. In general, if $\sum_{j=0}^{n} \theta_j \, x^j$, we can define n variables $z_j = x^j$, $.j = 1 \cdots n$, and convert the original equation to

$$y = \sum_{j=0}^{n} \theta_j \, z^j \tag{2.35}$$

and solve for a multiple linear regression linear fit.

TABLE 2.4: Heat Capacity of
Hydrogen Bromide and Temperature
in Kelvin

Capacity	Temp
10.79	118.99
10.80	120.76
10.86	122.71
10.93	125.48
10.99	127.31
10.96	130.06
10.98	132.41
11.03	135.89
11.08	139.02
11.10	140.25
11.19	145.61
11.25	153.45
11.40	158.03
11.61	162.72
11.69	167.67
11.91	172.86
12.07	177.52
12.32	182.09

2.4.1 Polynomial Regression Using R

This time we are going to use a dataset (see Table 2.4) that is called the Heat Capacity dataset obtained from the University of Florida website on regression data. This dataset records the heat capacity in cal/kmol and temperature in Kelvin degrees for solid hydrogen bromide in 18 experimental runs [12].

Below is an R script to compare three regression models: one linear, one quadratic and the last one cubic in the independent variable, which is Capacity in this case.

```
1  #Taking the Heat Capacity dataset from University of Florida Regression Datasets
2  #and performing linear and polynomial regression
3  #Load library for boxplotting
4  library(gplots)
5
6  HeatCapacity <- read.csv("~/Datasets/txt/HeatCapacity.txt", comment.char="#")
7  View(HeatCapacity)
8  plot(HeatCapacity)
9  Capacity <- HeatCapacity$Capacity
10 Temp <- HeatCapacity$Temp
11
12 #Model 1, one independent variable
13 model1 <- lm (Temp ~ Capacity, data=HeatCapacity)
14 summary(model1)
15 predictedVals1 <- predict.lm(model1)
16 lines(Capacity, predictedVals1, col="blue", lty="solid")
```

```
17
18  #Model 2, two independent variables
19  Capacity2 <- Capacity^2
20  model2 <- lm (Temp ~ Capacity + Capacity2,  data=HeatCapacity)
21  summary(model2)
22  predictedVals2 <- predict.lm (model2)
23  lines(Capacity, predictedVals2, col="red", lty="dashed" )
24
25  #Model 3, three independent variables
26  Capacity3 <- Capacity^3
27  model3 <- lm (Temp ~ Capacity + Capacity2 + Capacity3 , data=HeatCapacity)
28  summary(model3)
29  predictedVals3 <- predict.lm (model3)
30  lines(Capacity, predictedVals3, col="red", lty="dotted" )
31
32  #view the residuals and boxplot them
33  res1 <- residuals (model1)
34  res2 <- residuals (model2)
35  res3 <- residuals (model3)
36  #make a matrix out of the three residual vectors
37  resMat <- cbind (res1, res2, res3)
38  #do a boxplot
39  boxplot(resMat)
40
41  #Perform ANOVA comparison of the three models
42  anova(model1, model2, model3)
```

Selected lines of code are explained below. The dataset is read from file and stored in the variable called HeatCapacity. It can be viewed and plotted. The plot is given in Figure 2.6. First, we perform a linear least squares regression and print a summary of the fitted model called model1. By inspecting the plot, we can surmise that a linear regression may fit reasonably well, although

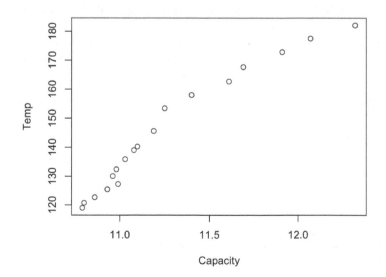

FIGURE 2.6: Scatter Plot of the Heat Capacity dataset

a quadratic fit may be better. It fits an LSRL to `Temp` given `HeatCapacity`. The summary is shown below.

```
Coefficients:
             Estimate Std. Error t value Pr(>|t|)
(Intercept) -348.277     29.576  -11.78 2.70e-09 ***
Capacity      43.761      2.621   16.70 1.52e-11 ***
---
Signif. codes: 0 '***' 0.001 '**' 0.01 '*' 0.05 '.' 0.1 ' ' 1

Residual standard error: 4.961 on 16 degrees of freedom
Multiple R-squared:  0.9457,       Adjusted R-squared:  0.9423
F-statistic: 278.8 on 1 and 16 DF,  p-value: 1.516e-11
```

The significance codes for both the `Intercept` and the coefficient for `Capacity` are ***, so both parameters are reliably obtained. Next, we use `model1` to predict the $\hat{y}$ values for the x values in the training data using the linear model. We also draw the regression line. This is the straight line in Figure 2.8.

```
predictedVals1 <- predict.lm(model1)
lines(Capacity, predictedVals1, col="blue", lty="solid")
```

In R, we can obtain the quadratic fit in terms of the independent variable `Capacity` (x) by creating a new variable `Capacity2`, which is the square of `Capacity`, and so x^2, and then fitting a model using multiple linear regression in two variables x and x^2.

```
Capacity2 <- Capacity^2
model2 <- lm (Temp ~ Capacity + Capacity2,  data=HeatCapacity)
```

There are other ways to obtain a quadratic fit, but this is a valid way as we have seen in the theoretical discussions. In this approach, we use R to perform multi-dimensional linear regression in x and x^2, and thereby actually perform non-linear regression or polynomial regression in this case. We summarize the model below.

```
Coefficients:
             Estimate Std. Error t value Pr(>|t|)
(Intercept) -3403.875    408.002  -8.343 5.12e-07 ***
Capacity      576.460     71.098   8.108 7.30e-07 ***
Capacity2     -23.174      3.092  -7.494 1.91e-06 ***
---
Signif. codes: 0 '***' 0.001 '**' 0.01 '*' 0.05 '.'  0.1 ' ' 1

Residual standard error: 2.353 on 15 degrees of freedom
Multiple R-squared:  0.9886,       Adjusted R-squared:  0.987
F-statistic:  648 on 2 and 15 DF,  p-value: 2.747e-15
```

The significance codes for `Intercept`, `Capacity` and `Capacity`^{2} are ***, indicating that R can find the parameters in a reliable manner. Finally, we obtain a cubic fit to the dataset by performing multi-dimensional linear regression in x, x^2 and x^3 as the three dimensions. We plot the cubic fit as well. The summary of the cubic fit is given below.

```
Coefficients:
               Estimate Std. Error t value Pr(>|t|)
(Intercept) -12993.483   10715.321  -1.213    0.245
Capacity      3078.252    2794.337   1.102    0.289
Capacity2     -240.513     242.694  -0.991    0.338
Capacity3        6.287       7.020   0.896    0.386

Residual standard error: 2.368 on 14 degrees of freedom
Multiple R-squared:  0.9892,        Adjusted R-squared:  0.9869
F-statistic: 426.5 on 3 and 14 DF,  p-value: 5.437e-14
```

Note that the significance codes for the coefficients are not good. So, even though the cubic model fits the data well, the coefficients are not computed in a stable way. In other words, R has not been able to determine the values of the coefficients in a reliable manner, e.g., there may be different combinations of coefficient values that may provide the same level of fit, but it is not able to determine good stable values.

The three regression models that we fit are given below.

$$t = -348.277 + 43.761x \qquad (2.36)$$

$$t = -3403.875 + 576.460x - 23.174x^2 \qquad (2.37)$$

$$t = -12993.483 + 3078.252x - 240.513x^2 + 6.287x^3 \qquad (2.38)$$

where $x =$ `Capacity`. We find that all three overall regression models are statistically valid with very small p-values of 1.516×10^{-11}, 2.747×10^{-15} and 5.437×10^{-14}, respectively. The R^2 values are 0.9457, 0.9886 and 0.9892 for the three fits, saying that cubic fit explains 98.92% of the variance in the dataset, only very slightly better than the quadratic fit. The R^2_{adj} values are 0.9423, 0.987 and 0.9869 for the three cases, making the quadratic and cubic fits indistinguishable in terms of goodness of fit. As a result, we can use either the quadratic or cubic fit in this case, since the difference in goodness is miniscule. It is clear from Figure 2.8 as well, where the quadratic and cubic fits are right on top of each other. In this figure, the quadratic fit is dashed black, and the cubic fit is red dotted. However, the coefficients of the cubic model are not very reliable, as noted before. Hence, the quadratic fit is the one that should be used.

Since we have three different regression models, we may want to visualize their quality in another way as well. This way involves plotting the residuals using what is called a bloxplot. A boxplot for a set of values shows how the values are distributed in terms of five simple statistics that can be obtained after we have sorted the values. They are the minimum value in the set, the

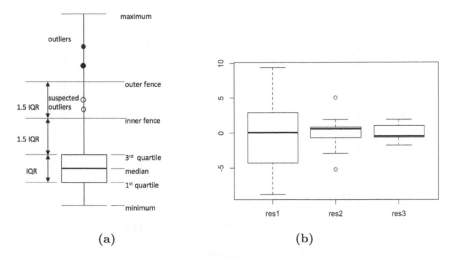

(a)   (b)

FIGURE 2.7: (a) A generic boxplot showing terminologies used. (b) Boxplots for the three regression fits: linear, quadratic and cubic.

maximum value in the set, the median value that lies in the middle location in the sorted sequence, the first quartile value that sits halfway between the smallest and the median, and the third quartile value that sits halfway between the median and the maximum value in the sorted sequence. A generic boxplot is shown in 2.7(a). The difference between the value of the first quartile and the value of third quartile is called *Inter Quartile Range* or IQR. A boxplot also creates what are called fences, an inner fence at a distance $1.5 \times IQR$, and an outer fence at a distance $3 \times IQR$. Points that lie in between the inner fence and the outer fence are called suspected outliers; points that lie in between the outer fence and the maximum are termed outliers. There are many other ways to compute outliers, but boxplots use this simple technique to give a quick feel of the spread of the values in a dataset. The same is done in the lower part of the diagram also, although not shown here. The boxplots for the three regression fits we are discussing are shown in 2.7(b). It shows that the residuals are much larger in magnitude and more spread out for the linear fit, indicating a worse fit than the other two, for which the spread of residuals is more or less similar, although the locations of the medians are different. The residuals for the quadratic fit has two "suspected" outliers.

We also perform a pairwise ANOVA comparison of the three models. The comparison is given below.

```
Analysis of Variance Table

Model 1: Temp ~ Capacity
Model 2: Temp ~ Capacity + Capacity2
```

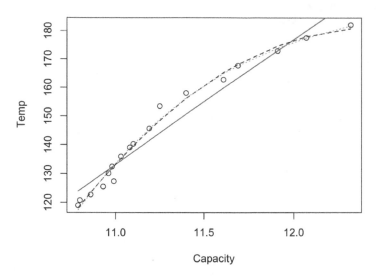

FIGURE 2.8: Plot of three Polynomial Fits on the Heat Capacity dataset

```
Model 3: Temp ~ Capacity + Capacity2 + Capacity3
   Res.Df     RSS Df Sum of Sq       F   Pr(>F)
1      16 393.86
2      15  83.03  1   310.827 55.4131 3.135e-06 ***
3      14  78.53  1     4.499  0.8021    0.3856
---
Signif. codes:  0 '***' 0.001 '**' 0.01 '*' 0.05 '.' 0.1 ' ' 1
```

ANOVA or Analysis of Variance is another test of significance we can perform. ANOVA can be used to compare several hypotheses, here regression fits, either all together or pairwise. In general, it can be used to compare the population means of several data collections. For example, here we have three regression fits, and corresponding to each fit we have a set of residuals. For ANOVA to work, just like all the regression models and significance tests (F-test and t-test) we have discussed before, the data collections (here, the three sets of residuals) must be distributed normally, i.e., using the Gaussian distribution. The way we perform the ANOVA test here is pairwise. We compare model1 and model2 first. Line 2 of the output above says that there is significant difference between model1 and model2 because the p-value is 3.13×10^{-6}. Next, we compare model2 with model3. Line 3 says that there is not a significant difference between model2 and model3 since the p-value is 0.3856 and it is much bigger than 0.05, the significance threshold. Thus, our pairwise testing says model2 and model3 are significantly different from model1. There are other ways to run ANOVA as well, and we will discuss it in an appropriate part of the book later.

2.5 Overfitting in Regression

Overfitting happens when a trained model fits very well to the training data, but the trained model does not do so well when used for prediction on data that were not used for training. Overfitting may take place if there are coincidental or accidental patterns in the training data. Such accidental patterns could arise due to the presence of noise in the data, or because the size of the training dataset is small, or can happen just randomly. This can happen in regression because the sample used to obtain the fit could be small or may have linearities or other relationships that would not occur in a bigger dataset. Incidental patterns that occur in the training data are unlikely to occur in general, i.e., in previously unseen data, on which the trained model may be used.

One of the main purposes of a machine learning algorithm is to be able to generalize, i.e., look past such accidental or random local patterns and be able to perform well with previously unseen data. In the case of regression, it would mean that we would like to be able to predict well the value of the dependent variable for previously unseen examples after obtaining a model trained on a dataset. A regressor that fits very well to the data, i.e., has very low sum of squared errors, may in fact, be a bad generalizer; we say that such a regressor has *overfit* the data. If a regressor fits very well to the data, it means that if there are errors in the data, it fits well to the error as well. A regressor that does not fit at all well to the data, i.e., has very high sum of squared errors, is also bad; we say that such a regressor *underfits* the data. The goal in machine learning, and regression in particular, is to neither underfit nor overfit.

2.6 Reducing Overfitting in Regression: Regularization

We want to reduce overfitting in regression so that it is able to reduce the effects of noise and generalize well. There are several approaches to reduce overfitting. Common methods include fitting splines and use what is called regularization. We will discuss a few regularization methods in this section.

When we obtain a fitted function using ordinary least squares linear regression, if the number of independent variables is small, the coefficients are easily understood and as a result, the fit is easy to interpret. However, if we have many independent variables, and an independent variable is linearly related strongly to another independent variable, these variables both may still occur in the fitted equation, making it unnecessarily complex. Even if a predictor is unrelated to the dependent variable, the corresponding coefficient may not be zero.

Thus, a preferred solution to the regression problem may be that we want to fit a linear (or maybe, even higher order) function, but dampen the sizes of the coefficients in a controlled manner, so that the fit itself may be controlled as desired. Such methods are called *shrinkage* methods. Shrinkage methods suppress the values of the coefficients, preventing the fitted equation from overfitting to the training dataset, i.e., providing stability to the predictions. Some shrinkage methods can also reduce certain coefficients to zero, performing a type of feature reduction or selection. In other words, such a regression method may remove features that do not affect the dependent variable strongly or do so only incidentally, i.e., in this training dataset only. It is also possible that if there are two features that are related to each other, it obtains the relationship more accurately, possibly by removing one of them from consideration, by amalgamating the effect of the two into one.

We discuss three methods of shrinkage regression here: Ridge regression, Lasso regression and Elastic Net regression.

2.6.1   Ridge Regression

Ridge Regression attempts to dampen the coefficients of a linear regression fit. Ridge regression adds an extra component to the error or loss function. The objective of the optimization function becomes

$$L = \sum_{i}^{N} \left(y^{(i)} - \theta_0 - \sum_{j=1}^{n} \theta_i x_j^{(i)} \right)^2 + \lambda \sum_{j=1}^{n} \theta_j^2 \qquad (2.39)$$

$$= LSE + \lambda \sum_{j=1}^{n} \theta_j^2 \qquad (2.40)$$

where LSE is the Least Squared Error we have discussed before. The term $\lambda \sum_{j=1}^{n} \theta_j^2$ is called the shrinkage penalty. Since it uses the squares of the coefficients, it is also called an L_2-loss. Such an optimization objective has the effect of shrinking the sizes of the coefficients toward 0, assuming λ is positive. The values of the θ_is can be positive or negative.

The constant λ is a tuning parameter that controls or regulates the amount of shrinkage of the regression coefficients. For example, if $\lambda = 0$, there is no shrinkage; it is simply the linear least squares regression. If $\lambda \to \infty$, the values of the θ_js can be made arbitrarily small. Thus, the fitted equations that come out of Ridge Regression are not unique and depend on the value of λ, and it is important to use good values of λ. This is usually done using the approach called cross-validation.

The parameter θ_0 is the intercept on the Y-axis of the model. If θ_0 is shrunk or penalized, we may need or force the model to always have low intercept. Thus, it is recommended that θ_0 be estimated separately. A recommended values for the intercept is

$$\theta_0 = \bar{y} = \frac{1}{N} \sum_{i=1}^{N} y^{(i)}.$$

The remaining parameters are then obtained using an optimization technique.

To be able to use ridge regression in R, we use a library called *glmnet*[2]. It provides implementations for a few general linear regression models including ridge regression. We provide a simple program in R below and then discuss the steps. The code is commented well and divided into groups separated by blank lines.

```
1  #glmnet implements General Linear Models for regression
2  library(glmnet)
3
4  #Read the CSV dataset
5  Immigrants <- read.csv("~/Datasets/txt/Immigrants.csv")
6  View(Immigrants)
7  plot(Immigrants)
8
9  #To be able to use glm libary, one needs to convert the data into a matrix
10 immigrantData <- as.matrix (Immigrants)
11
12 #trainX is all columns but the last column of the entire dataset
13 trainX <- immigrantData [,1:ncol(immigrantData)-1]
14 #trainY is the last column of the entire dataset
15 trainY <- immigrantData [,ncol(immigrantData)]
16
17 #Perform regression, alpha=0 perform ridge regression.
18 #The call to glmnet tries a lot of values of lambda, from large values to small.
19 #Do not use one value of lambda, the library tries values of lambda and plots.
20 fit1 <- glmnet(trainX,trainY, alpha=0)
21
22 #Plot the variables against SSE: norm, dev and lambda
23 plot (fit1)
24 #Plot the coefficients
25 coefficients (fit1)  #Show coefficients for fit with various values of lambda
26
27 #Shows how the squared error (SSE) changes as the value of lambda changes.
28 #One needs to obtain a value of lambda that works.
29 #One can do so by inspection:
30 plot (fit1, xvar="lambda")
31
32 #Find the value of lambda that gives the best fit using cross validation
33 findLambdas <- cv.glmnet(x = trainX,y = trainY,alpha = 0)
34 #Plot how the SSE changes as we change the value of lambda
35 plot (findLambdas)
36
37 #Authors of glmnet recommend use of  lambda that is 1 standard dev away
38 fittedModel1 <- glmnet(trainX,trainY,lambda=findLambdas$lambda.1se)
39
40 #Fit the model for the value of lambda that gives the min SSE
41 fittedModel2 <- glmnet(trainX,trainY,lambda=findLambdas$lambda.min)
42
43 #Create a test example, and make it a row vector
44 #English Proficiency = 10, Literacy = 60 and Residency = 80
45 testX <- as.matrix (c (10, 60, 80))
46 testX <- t (testX) #transpose the matrix
47
48 #Predict Income for testX using the two models
```

```
49  predict (fittedModel1, newx=testX)
50  predict (fittedModel2, newx=testX)
51
52  #Predict Income for three cases: (10, 60, 80), (70, 80, 90), and (80, 90, 100)
53  testX = matrix ( c(10, 60, 80,  70, 80, 90, 80, 90, 100), nrow=3, ncol=3)
54  predict (fittedModel1, newx=testX)
55
56  predict (fittedModel2, newx=testX)
```

The first line of code simply says that we use the `glmnet` library. The `library()` command loads the package and any add-on packages and makes it available in the program. The program then reads a CSV file and makes it available in the program as a data frame called `Immigrants`. `glmnet` requires the dataset to be available as a matrix. R provides a function called `as.matrix()` to make this conversion since this is a common need.

```
ImmigrantData <- as.matrix (Immigrants)
```

Next, we take the all but the last column of the matrix, i.e., the first three columns as the predictor variables for the entire training set, and call them `trainX`. The corresponding predictor variables are in the single-column matrix or a vector called `trainY`.

```
#trainX is all columns but the last column of the entire dataset
trainX <- immigrantData [,1:ncol(immigrantData)-1]
#trainY is the last column of the entire dataset
trainY <- immigrantData [,ncol(immigrantData)]
```

Once we have the training set in terms of the matrix `trainX` and the vector `trainY`, we can run ridge regression.

```
fit1=glmnet(trainX,trainY, alpha=0)
```

In `glmnet`, the value of the parameter `alpha` is set to 0 for ridge regression. As we know from our discussions before, ridge regression has a parameter called λ. `glmnet` iterates over values of λ and obtains the values of the linear coefficients for the three predictor variables for each value of λ. We can look at the values of the coefficients as the value of λ changes. We view the coefficients by using the statement given below.

```
coefficients (fit1)
```

There are three coefficients and an intercept, thus four values for each value of λ. It tries 100 values of λ and generates the coefficients for each value of λ.

```
4 x 100 sparse Matrix of class "dgCMatrix"
   [[ suppressing 100 column names 's0', 's1', 's2' ... ]]

(Intercept) 5.520286e+01 54.890778349 54.860457034 54.827201014
English     8.946241e-36  0.010326864  0.011330168  0.012430568
Literacy    9.755401e-37  0.001126902  0.001236473  0.001356665
Residency   1.244163e-36  0.001435737  0.001575179  0.001728107
```

```
(Intercept) 54.790728558 54.750731346 54.706872070 54.658781817
English      0.013637386  0.014960818  0.016412023  0.018003201
Literacy     0.001488503  0.001633105  0.001791700  0.001965629
Residency    0.001895813  0.002079711  0.002281347  0.002502412
.
.
.
(Intercept) -6.959630548 -6.15360465 -5.3613479
English      1.231559574  1.15821787  1.0854865
Literacy     0.793924662  0.80957092  0.8246066
Residency   -0.009028241 -0.01964026 -0.0297369
```

We can also plot the values of the coefficients as the value of λ changes. The values of λ are not plotted directly, but in terms of their log; this allows for a much larger range of λ values to be considered. We obtain this graph by using the command:

```
plot (fit1, xvar="lambda")
```

in glmnet. Figure 2.9 shows a graph with coefficients on the Y-axis and λ values on the X-axis. It shows the coefficients as y values that we would have obtained for regular least squares linear regression when $log(\lambda)$ is near the origin. It also shows that values of two of the coefficients first grow and then fall toward 0 as λ increases. For $\lambda \approx 1000$, all the coefficients become very

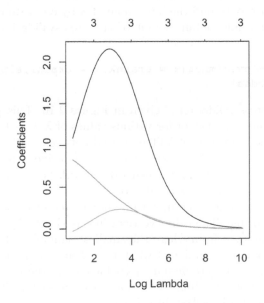

FIGURE 2.9: Values of the coefficients as λ changes

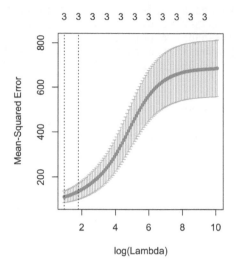

FIGURE 2.10: Ridge Regression cross-validation: Values of the coefficients as λ changes

low, close to 0. The value of the third coefficient is the highest for the smallest value of λ, and then falls all the way as the values of λ become bigger.

Now the question arises: Which value of λ should one use? The `glmnet` library provides a way to find the best value of λ by cross-validation. We can run cross-validation and obtain a plot of the cross-validation by using the following.

```
findLambdas <- cv.glmnet(x = trainX,y = trainY,alpha = 0)
plot (findLambdas)
```

The plot of the cross-validation is given in Figure 2.10. This plot shows the values of the mean squared error for various values of λ. log of λ is plotted on the X-axis. The actual values of λ fitted varied between $log(0)$ which is 1 and $log(2^{10})$, which is $1,204$, in this case. The package decides what values to try, but it tries a lot of values to find which values work better. It also plots the values so that the user can pick a value of λ if need be. The plot also shows the value of λ (λ_{min}) where the best value of squared error occurs. It occurs on the left vertical line. The other vertical line is the λ (λ_{1se}) value at one standard deviation away from where the smallest sum of least squared errors occurs. The plot has the number 3 written on top several times, indicating that there are 3 non-zero parameters or coefficients in the model that results as λ changes. That is, the number of non-zero parameters in always 3, no matter what value of λ is picked, in this case.

The authors of `glmnet` recommend that the value of λ used is not the one that is one standard deviation away. In the code snippet given below, we also fit a model with the λ value at the point where the lowest squared error occurs. The exact value of λ is left to the user.

```
#Authors of glmnet recommend use of lambda that is 1 standard deviation away
fittedModel1 <- glmnet(trainX,trainY,lambda=findLambdas$lambda.1se)

#Fit the model for the value of lambda that gives the min SSE
fittedModel2 <- glmnet(trainX,trainY,lambda=findLambdas$lambda.min)
```

In this case, the least value of the sum of squared errors occurs at `lambda.min` = 2.611931. Following the advice of the implementors of `glmnet`, let us choose the value of λ to be `lambda.1se` = 5.009449, a standard deviation away from 2.611931. This assumes that the squared errors or residuals are normally distributed, as we have always assumed in the discussion of regression in this chapter. Once we have fit these two models, we can find the values that they predicted at various data points for the two values of λ.

```
Create a test example, and make it a row vector
#English Proficiency = 10, Literacy = 60 and Residency = 80
testX = as.matrix (c (10, 60, 80))
testX = t (testX)

#Predict Income for testX using the two models
predict (fittedModel1, newx=testX)
predict (fittedModel2, newx=testX)
```

This produces 54.38512 and 54.22761 as the predicted values using the two models. We can find the predicted for several values or a matrix of values at the same time.

```
#Predict Income for three cases: (10, 60, 80), (70, 80, 90), and (80, 90, 100)
testX = matrix ( c(10, 60, 80,  70, 80, 90, 80, 90, 100), nrow=3, ncol=3)
```

The predicted outputs for the two cases are given below,

```
> predict (fittedModel1, newx=testX)
          s0
[1,] 61.59436
[2,] 68.80359
[3,] 76.01282
> predict (fittedModel2, newx=testX)
          s0
[1,] 62.82551
[2,] 71.42341
[3,] 80.02131
```

To find the values of the coefficients at a specific value of λ, we can use the `coeff ()` statement in `glmnet`. Writing

```
coeff (fit1, s=2.611931)
```

gives the values of the coefficients at the λ_{min}.

```
(Intercept) 2.6402144
English       .
Literacy    0.8597899
Residency     .
```

The . (dot) means the value is too small to be written out. So, the fitted equation is

$$\hat{y} = 0.8597899 x_2 + 2.6402144. \tag{2.41}$$

The fitted equation at λ_{1se} is

$$\hat{y} = 0.7624963 x_2 + 8.5881932. \tag{2.42}$$

Here, $y = $ Income, and $x_2 = $ Literacy. The other two predictor variables have negligible influence on Income, although the cross-validation graph tells us that they were not eliminated.

2.6.2　Lasso Regression

In Ridge Regression, we see that as the value of the hyperparameter λ becomes bigger, the parameters $\theta_0 \cdots \theta_n$ become smaller and smaller, but the way it has been designed the parameters never become 0 although they could become quite small. Thus, all the features are likely to matter in the final regression equation we come up with no matter what value of λ we choose, although some may become marginal if the corresponding coefficient becomes really small in absolute value. The fact that all independent variables remain at the end, may not matter much in this example with three independent variables or features, but if we have a lot of features to begin with, say tens or hundreds or even more, Ridge Regression will keep them even though the coefficients may become tiny. Making some of the coefficients exactly 0 so that they do not matter at all in the final equation may improve interpretability of the regression equation since the number of variables will become smaller. For example, if there were 50 independent variables to begin with and we are left with only 10 at the end, it becomes much easier to visualize or understand the relationships between the ten independent variables and the dependent variable. Lasso Regression has been designed to achieve this type of reduction in the number of independent variables that matter as λ becomes bigger. The effect of removing independent variables from consideration is called feature selection.

Lasso Regression is quite similar to Ridge Regression in formulation, but instead of an L_2-loss in Ridge, it uses absolute value regression.

$$L = \sum_{i}^{N} \left(y^{(i)} - \theta_0 - \sum_{j=1}^{n} \theta_i x_j^{(i)} \right)^2 + \lambda \sum_{j=1}^{n} \left| \theta_j \right| \tag{2.43}$$

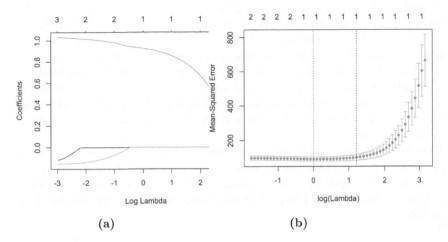

FIGURE 2.11: (a) Values of the coefficients as λ changes in Lasso Regression, (b) Mean Squared Error as λ changes in Lasso Regression

$$= LSE + \lambda \sum_{j=1}^{n} \left| \theta_j \right| \tag{2.44}$$

Here λ is a hyperparameter, and like in Ridge Regression, it is used to increase the prediction accuracy of regression in addition to providing better interpretability. For our example with three dependent variables, Figure 2.11 shows how the coefficients (θ_0, θ_1 and θ_2) vary as $log(\lambda)$ changes from -3 to slightly over 3. When $log(\lambda) = -3$ or $\lambda = e^{-3} = \frac{1}{e^3} = \frac{1}{20.086} = 0.0498$, all three coefficients have non-zero values. When $log(\lambda) = 0$ or $\lambda = 1$, only one coefficient is non-zero. When $log(\lambda) = 3$ or $\lambda = e^8 = 20.086$, all coefficients are forced to 0, or that the regression is trivial $\hat{y} = 0$.

2.6.3 Elastic Net Regression

We have seen that Ridge Regression pushes the values of the coefficients toward 0, but does not make them exactly 0, as the hyperparameter λ rises in value. We have also seen that Lasso Regression pushes the values of coefficients to 0 as the value of λ becomes large, eliminating features as λ rises. Thus, we see that both regression techniques produce a family of learned models, not just one. Proper choice of λ that leads to the regression equation we finally use, needs search through cross validation. The authors of Ridge and Lasso regression software in `glmnet` recommend using a value where RMS error is the lowest, and also alternatively, a value of λ one standard deviation away in the positive direction of λ. The ultimate choice is the user's based on real-world considerations outside of mathematics.

A problem both ridge and lasso regressions try to solve is how to handle features that are related to each other linearly. It has been observed that if there are two collinear or linearly related features, the coefficient of one may become zero much before the other does in Lasso Regression, and thus, some similar features are left in whereas others are pushed out of consideration. Thus, related features are not treated similarly, and why this happens to which feature is not systematic or easily explainable. Ridge regression pushes coefficients of all of the related features down in a more predictable manner, although not down to 0.

Elastic Net Regression attempts to combine the effects of both Lasso and Ridge regressions. In other words, Elastic Net Regression treats related features similarly as coefficients are shrunk, and also helps in feature selection. The objective function for Elastic Net Regression minimizes the loss function given below.

$$L = \sum_{i}^{N} \left(y^{(i)} - \theta_0 - \sum_{j=1}^{n} \theta_i x_j^{(i)} \right)^2 + \lambda \left((1 - \alpha) \sum_{j=1}^{n} \theta_j^2 + \alpha \sum_{j=1}^{n} |\theta_j| \right) \quad (2.45)$$

$$= LSE + \lambda \left((1 - \alpha) \sum_{j=1}^{n} \theta_j^2 + \alpha \sum_{j=1}^{n} |\theta_j| \right) \quad (2.46)$$

Thus, the two losses, one used in Ridge and the other used in Lasso, are both used, and which one is predominant depends on the value of α. We have the λ hyperparameter as before to exert the chosen amount of shrinkage pressure on the coefficients used in the regression. Ideally, we should do a search over possible values of both λ and α to find the combination that works best. Thus, a grid search or some intelligent search in both parameters may be helpful. Note that in `glmnet`, if `alpha` is set to 0, it means we are performing Ridge Regression, and if `alpha` is set to 1, it means that we perform Lasso Regression. Any other value of `alpha` makes it Elastic Net Regression.

In the run whose results we report here, we set $\alpha = 0.5$ so that both constraints, Ridge and Lasso, are equally weighted. `glmnet` will perform cross validation to find the right value of λ that minimizes the sum of squared errors for a given value of α. Figure 2.12(a) shows the search over values of $log(\lambda)$, and Figure 2.12(b) shows the change in mean squared error as $log(\lambda)$ changes.

2.7   Regression in Matrix Form

So far, we have discussed regression using a detailed element-wise "calculus" view. We wrote algebraic expressions for the sum of squared

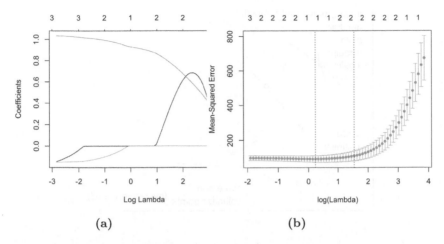

FIGURE 2.12: (a) Values of the coefficients as λ changes in Elastic Net Regression, (b) Mean Squared Error as λ changes in Elastic Net Regression

errors or distances between each observed point in the training dataset and the regression line or function. The expression was then minimized by obtaining the partial derivatives, setting each partial derivative to 0, obtaining a system of linear equations, and performing algebra to obtain the regression coefficients, $\theta_0, \theta_1, \cdots, \theta_n$. We walked through the steps and then later relied on the statistical package R to solve regression problems without really knowing how they are programmed internally.

This process requires copious algebra as we have seen. It is perfectly reasonable to do so, but in this section, we will first present a more compact and more elegant view, using geometry and linear algebra, of how the same problem of obtaining regression coefficients can be solved. Later, we will look at the same problem using matrix calculus. The reason for discussing the matrix formulation is to impress upon the reader that actual computational implementations, especially for solving problems with a large number of variables is likely to be vector and matrix-based, so that they can be executed on fast GPUs (Graphical Processing Units).

2.7.1 Deriving LSRL Formula using Geometry and Linear Algebra

Suppose we are given two points, $\langle x_1, y_1 \rangle$ and $\langle x_2, y_2 \rangle$ in 2-D space. Unless the points are exactly the same, we can draw a straight line joining them. If the equation of the line is $y = \theta_0 + \theta_1 x$, we can write the following two

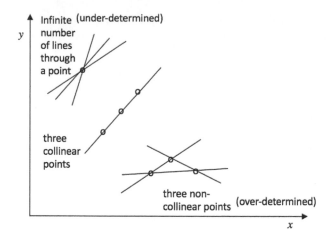

FIGURE 2.13: Number of lines that can be drawn through several points

equations

$$y_1 = \theta_0 + \theta_1 x_1$$
$$y_2 = \theta_0 + \theta_1 x_2$$

and find the two parameters θ_0 and θ_1 for the line by solving the system of two linear equations in two unknowns.

We can rewrite the system of these two equations more compactly in matrix form. If

$$\mathbf{A} = \begin{bmatrix} 1 & x_1 \\ 1 & x_2 \end{bmatrix} \quad \vec{\theta} = \begin{bmatrix} \theta_0 \\ \theta_1 \end{bmatrix} \quad \vec{y} = \begin{bmatrix} y_1 \\ y_2 \end{bmatrix}, \tag{2.47}$$

the set of linear equations can be written simply as

$$\mathbf{A}\,\vec{\theta} = \vec{y}. \tag{2.48}$$

This is a matrix equation. If in the 2-D case, we are given one point only, we can draw an infinite number of lines through it, and it is an under-determined system. If we are given three or more distinct points instead of two, and the points are not collinear, we cannot a draw a single line that goes through all of them. This is the case of an over-determined linear system. In such a case, we have to draw a line that "best" fits the three points—this is regression. These situations are shown in Figure 2.13.

For example, in our Drugs and Math dataset, we have two dimensions, but we are given seven points. The values of $\mathbf{A}$, $\vec{\theta}$ and $\vec{y}$ are given below for this

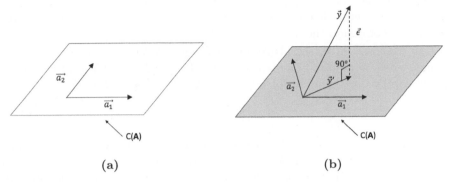

FIGURE 2.14: (a) Column Space of a Matrix **A**, (b) Column Space of a Matrix **A** with Projection

dataset.

$$\mathbf{A} = \begin{bmatrix} 1 & 78.93 \\ 1 & 58.20 \\ 1 & 67.47 \\ 1 & 37.47 \\ 1 & 45.65 \\ 1 & 32.92 \\ 1 & 29.97 \end{bmatrix} \quad \vec{\theta} = \begin{bmatrix} \theta_0 \\ \theta_1 \end{bmatrix} \quad \vec{y} = \begin{bmatrix} 1.17 \\ 2.97 \\ 3.26 \\ 4.69 \\ 5.83 \\ 6.00 \\ 6.41 \end{bmatrix} \tag{2.49}$$

Clearly, we have two variables only, but a system of seven equations in these two variables. Thus, this is an over-determined system, where there is no exact solution to it. In this case, it is worthwhile to try to find a solution such that

$$\mathbf{A}\,\vec{\theta} \approx \vec{y}, \tag{2.50}$$

i.e., $\mathbf{A}\,\vec{\theta}$ is as close to (i.e., $\approx \vec{y}$) as possible.

Another concept that is useful is the column space $C(\mathbf{A})$ of any matrix **A**. $C(\mathbf{A})$ is a plane which contains all linear combinations of the columns of **A**. In other words, the vectors corresponding to the columns of **A** lie in $C(\mathbf{A})$ as well. If **A** is 2×2 like the one in Equation 2.47, and the columns of **A** are called $\vec{a_1}$ and $\vec{a_2}$, all linear combinations of them form $C(\mathbf{A})$, and as a result, $\vec{a_1}$ and $\vec{a_2}$ lie in $C(\mathbf{A})$ also. These two vectors are shown in Figure 2.14(a). Any solution to $\mathbf{A}\,\vec{\theta} = \vec{y}$ must be on this plane.

However, if the matrix equation is over-determined as in Equation 2.49, there is no exact solution to the system. In this case, the observed values of y, which are given in $\vec{y}$, cannot lie in $C(\mathbf{A})$. In other words, this $\vec{y}$ is outside $C(\mathbf{A})$. Without loss of generality, let us assume $\vec{y}$ sticks out like shown in Figure 2.14(b). We draw $\vec{a_1}$ and $\vec{a_2}$ as meeting at a point on $C(\mathbf{A})$, and $\vec{y}$, which is a vector in a higher dimensional space is meeting at that point as well. When we try to find a regression fit to $\vec{y}$, since it does not lie in $C(\mathbf{A})$, the best we can do is find a vector that lies in $C(\mathbf{A})$ and is as close to $\vec{y}$ as possible. A vector that is closest to $\vec{y}$ and is on $C(\mathbf{A})$ is the perpendicular projection of $\vec{y}$ onto the plane that is $C(\mathbf{A})$.

In Figure 2.14(b), the projection of $\vec{y}$ onto the plane that is $C(\mathbf{A})$ is $\vec{y'}$. Thus, when we want to fit a regression function to $\vec{y}$, we will fit it to $\vec{y'}$ instead. Therefore, we are going to solve for $\vec{\theta}$ to satisfy the equation

$$\mathbf{A}\,\vec{\theta} = \vec{y'} \tag{2.51}$$

instead of solving $\mathbf{A}\,\vec{\theta} = \vec{y}$. The vertical line from $\vec{y}$ to $C(\mathbf{A})$ is called $\vec{\epsilon}$, which is the vector of errors or residuals, one residual per observed point. Our goal is to make $\vec{\epsilon}$ as small as possible so that $\vec{y'}$ is as close to $\vec{y}$ as possible. As already mentioned, geometrically speaking, this is achieved by the orthogonal projection $\vec{y'}$.

Assume $\mathbf{A}$ has n columns and n rows, and has full rank, i.e., all rows are linearly independent. Let us call the ith column of $\mathbf{A}$ by $\vec{a_i}$. Then, we can write

$$\mathbf{A} = \left[\vec{a_1} \cdots \vec{a_i} \cdots \vec{a_n} \right] \tag{2.52}$$

where

$$\vec{a_i} = \begin{bmatrix} a_{1i} \\ \vdots \\ a_{ni} \end{bmatrix}. \tag{2.53}$$

$\mathbf{A}$ is a single row matrix where each element is a column vector. If we take the transpose of $\mathbf{A}$, we get a single column where each element is a row vector, the transpose of $\vec{a_i}$.

We can now write

$$\mathbf{A}^T \vec{\epsilon} = \begin{bmatrix} \vec{a_1} \\ \vdots \\ \vec{a_n} \end{bmatrix} \vec{\epsilon}$$

$$= \begin{bmatrix} \vec{a_1} \circ \vec{\epsilon} \\ \vdots \\ \vec{a_n} \circ \vec{\epsilon} \end{bmatrix}$$

$$= \begin{bmatrix} 0 \\ \vdots \\ 0 \end{bmatrix}$$

$$= \mathbf{0} \tag{2.54}$$

where $\circ$ stands for the inner product or dot product, which is the element by element product followed by summation. Each dot product is 0 because $\vec{\epsilon}$ is perpendicular to $C(\mathbf{A})$, and each $\vec{a_i}$ is in $C(\mathbf{A})$. $\mathbf{0}$ is a vector with all 0s.

Going back to our problem stated in Equation 2.51, we can write

$$\mathbf{A}\,\vec{\theta} = \vec{y}'$$
$$= \vec{y} - \vec{\epsilon}. \tag{2.55}$$

If we premultiply by $\mathbf{A}^T$, we can write

$$\mathbf{A}^T\left(\mathbf{A}\,\vec{\theta}\right) = \mathbf{A}^T\left(\vec{y} - \vec{\epsilon}\right)$$
$$= \mathbf{A}^T\,\vec{y} - \mathbf{A}^T\,\vec{\epsilon}$$
$$= \mathbf{A}^T\,\vec{y} - \mathbf{0}$$
$$= \mathbf{A}^T\,\vec{y}.$$

Finally, we can write

$$\mathbf{A}^T\mathbf{A}\,\vec{\theta} = \mathbf{A}^T\,\vec{y}, \ \ or$$
$$\vec{\theta} = \left[\mathbf{A}^T\mathbf{A}\right]^{-1}\mathbf{A}^T\,\vec{y}. \tag{2.56}$$

Thus, we have obtained the parameters for regression, $\theta_0, \cdots, \theta_n$ in terms of a matrix equation. Although obtaining a matrix's inverse is not straightfoward, matrix operations can be implemented in modern hardware like GPUs better than straight element-wise calculations on vectors and matrices. This is one of the reasons why we have discussed this matrix-based approach here. In addition, although it requires background on matrices, it makes the derivations of formulas much shorter and cleaner. Note that Equation 2.7.2 works no matter how big $\vec{\theta}$ is, i.e., how many parameters we have. It works in general, whether we have one independent variable or a hundred or a thousand. Using the element-wise approach we discussed earlier, dealing with more than three or four variables becomes a cumbersome algebraic problem.

2.7.2 Deriving LSLR Formula Using Matrix Calculus

As noted in the previous section, using element-wise partial derivatives and setting them to 0 and solving a system of linear equations, are cumbersome and error-prone. That is why we introduced a geometric derivation of the LSLR formula using linear algebra and matrices. We can derive the formula given in Equation 2.7.2 using calculus, but this time using matrices and vectors, making the derivation compact.

Let us start with Equation 2.29, which is repeated below for convenience.

$$\epsilon^{(i)} = y^{(i)} - \theta_0 - \theta_1 x_1^{(i)} - \cdots - \theta_j x_j^{(i)} - \cdots - \theta_n x_n^{(i)}.$$

It gives the error or residual at the ith point in the training dataset when we fit a regression line with n dependent variables $x_1, \cdots, x_n$, with y being the dependent variable. If

$$\mathbf{A} = \begin{bmatrix} 1 & x_1^{(1)} & \cdots & x_n^{(1)} \\ \vdots & \vdots & \vdots & \vdots \\ 1 & x_1^{(N)} & \cdots & x_n^{(N)} \end{bmatrix} \quad \vec{\theta} = \begin{bmatrix} \theta_0 \\ \vdots \\ \theta_n \end{bmatrix} \quad \vec{y} = \begin{bmatrix} y^{(1)} \\ \vdots \\ y^{(N)} \end{bmatrix} \quad \vec{\epsilon} = \begin{bmatrix} \epsilon^{(1)} \\ \vdots \\ \epsilon^{(N)} \end{bmatrix}, \quad (2.57)$$

using Equation 2.55, we can write

$$\vec{\epsilon} = \vec{y} - \mathbf{A}\,\vec{\theta}. \tag{2.58}$$

Our objective in LSLR is to minimize the sum of the squares of all residuals $\sum_{i=1}^{N} \left(\epsilon^{(i)}\right)^2$. In terms of vectors, we can write the sum of squares as $\vec{\epsilon}^T \vec{\epsilon}$. This works because

$$\vec{\epsilon}^T \vec{\epsilon} = \begin{bmatrix} \epsilon^{(1)} & \cdots & \epsilon^{(N)} \end{bmatrix} \begin{bmatrix} \epsilon^{(1)} \\ \vdots \\ \epsilon^{(N)} \end{bmatrix}$$

$$= \left(\epsilon^{(1)}\right)^2 + \cdots + \left(\epsilon^{(N)}\right)^2$$

Hence, we can write that our objective function to minimize is

$$\vec{\epsilon}^T \vec{\epsilon} = \left(\vec{y} - \mathbf{A}\vec{\theta}\right)^T \left(\vec{y} - \mathbf{A}\vec{\theta}\right)$$

$$= \left(\vec{y}^T - \left(\mathbf{A}\vec{\theta}\right)^T\right)\left(\vec{y} - \mathbf{A}\vec{\theta}\right) \quad \text{since } \left(\mathbf{A} - \mathbf{B}\right)^T = \mathbf{A}^T - \mathbf{B}^T$$

$$= \left(\vec{y}^T - \vec{\theta}^T \mathbf{A}^T\right)\left(\vec{y} - \mathbf{A}\vec{\theta}\right) \quad \text{since } \left(\mathbf{A}\mathbf{B}\right)^T = \mathbf{B}^T \mathbf{A}^T$$

$$= \vec{y}^T \vec{y} - \vec{y}^T \mathbf{A}\vec{\theta} - \vec{\theta}^T \mathbf{A}^T \vec{y} + \vec{\theta}^T \mathbf{A}^T \mathbf{A}\vec{\theta}$$

$$= \vec{y}^T \vec{y} - 2\vec{\theta}^T \mathbf{A}^T \vec{y} + \vec{\theta}^T \mathbf{A}^T \mathbf{A}\vec{\theta}$$

The last line is true because $\vec{y}^T \mathbf{A}\vec{\theta} = \vec{\theta}^T \mathbf{A}^T \vec{y}$. We can write out the detailed products and see that they produce the same scalar. In particular, if we consider the matrices in $\vec{y}^T \mathbf{A}\vec{\theta}$, they are $1 \times n, n \times n$ and $n \times 1$ in size, producing a scalar. In $\vec{\theta}^T \mathbf{A}^T \vec{y}$, the dimensions are $1 \times n, n \times n$, and $n \times 1$, producing a scalar as well.

To minimize $\vec{\epsilon}^T \vec{\epsilon}$, we need to differentiate it and set it to 0. So, we need to compute $\frac{\partial}{\partial \vec{\theta}}\left(\vec{\epsilon}^T \vec{\epsilon}\right)$, where $\vec{\epsilon}^T \vec{\epsilon}$ is a scalar, and $\vec{\theta}$ is a vector. Therefore, according to rules of matrix differentiation, we get as many components as we have in $\vec{\theta}$ in the derivative.

$$0 = \frac{\partial}{\partial \vec{\theta}}\left(\vec{\epsilon}^T \vec{\epsilon}\right)$$

$$= \frac{\partial}{\partial \vec{\theta}}\left(\vec{y}^T \vec{y} - 2\vec{\theta}^T \mathbf{A}^T \vec{y} + \vec{\theta}^T \mathbf{A}^T \mathbf{A}\vec{\theta}\right)$$

$$= \frac{\partial}{\partial \vec{\theta}}\left(-2\vec{\theta}^T \mathbf{A}^T \vec{y} + \vec{\theta}^T \mathbf{A}^T \mathbf{A}\vec{\theta}\right)$$

The last line works because $\vec{y}^T \vec{y}$ is the sum the sum of squares of all the $y^{(i)}$ values, and is a constant as far as $\vec{\theta}$ is concerned and hence the derivative of the first term is 0. Continuing, we get

$$= -2\frac{\partial}{\partial \vec{\theta}} \left(\vec{\theta}^T \mathbf{A}^T \vec{y} \right) + \frac{\partial}{\partial \vec{\theta}} \left(\vec{\theta}^T \mathbf{A}^T \mathbf{A} \vec{\theta} \right)$$

The first term being differentiated, $\vec{\theta}^T \mathbf{A}^T \vec{y}$ is a scalar because the dimensions concerned are $1 \times n, n \times n$ and $n \times 1$. When we differentiate it with respect to $\vec{\theta}$, we will get a vector with the same dimensions as $\vec{\theta}$. If we think of $\mathbf{A}^T \vec{y}$ as some vector $\vec{b}$, the first term is a component-wise product of the two vectors $\vec{\theta}$ and $\vec{b}$. When we differentiate it with respect to $\vec{\theta}$, we get $\vec{b}$ or $\mathbf{A}^T \vec{y}$ back. So, at this time, we have

$$= -2\mathbf{A}^T \vec{y} + \frac{\partial}{\partial \vec{\theta}} \left(\vec{\theta}^T \mathbf{A}^T \mathbf{A} \vec{\theta} \right)$$

The second term is also a scalar and following rules of matrix calculus, we can show that its derivative is $\mathbf{A}^T \mathbf{A} \vec{\theta}$. Discussioin of matrix calculus can be found in many books and webpages[3]. Finally, we get

$$0 = -2\mathbf{A}^T \vec{y} + 2\mathbf{A}^T \mathbf{A} \vec{\theta}$$

By performing some simple algebra, we get the formula we obtained earlier in Equation 2.7.2, which is repeated below.

$$\vec{\theta} = \left[\mathbf{A}^T \mathbf{A} \right]^{-1} \mathbf{A}^T \vec{y}.$$

The loss function for Lasso Regression can be written as

$$L = \left(\vec{y} - \mathbf{A} \vec{\theta} \right)^T \left(\vec{y} - \mathbf{A} \vec{\theta} \right) + \lambda \vec{\theta}^T \vec{\theta} \tag{2.59}$$

and it can be matrix-differentiated and set to zero and then solved to obtain

$$\vec{\theta} = \left[\mathbf{A}^T \mathbf{A} + \lambda^2 \mathbf{I} \right]^{-1} \mathbf{A}^T \vec{y}.$$

where $\mathbf{I}$ is an identity matrix. We do not show the steps here, but they are similar to how we obtained the matrix equation for Ridge Regression. This shows that Ridge Regression has a form which is quite similar to LSLR.

The loss function for Lasso Regression can be written as

$$L = \left(\vec{y} - \mathbf{A} \vec{\theta} \right)^T \left(\vec{y} - \mathbf{A} \vec{\theta} \right) + \lambda |\vec{\theta}| \tag{2.60}$$

---

[3]For example, see Proposition 9 in `atmos.washington.edu/~dennis/MatrixCalculus.pdf`

However, because it uses the absolute value, taking derivatives and coming up with a closed form is not straightforward. So, it is with Elastic Net Regression.

2.8   Conclusions and Further Reading

This chapter has discussed a number of approaches to fitting a function to a dataset so that the learned function or model can be used for predictive purposes. In this chapter, we have discussed how to fit linear functions of single as well as multiple independent variables. We also have discussed how to obtain polynomial fits of single independent variables.

Regression is a well-discussed topic in the field of statistics. Numerous methods to perform linear and non-linear regression have been proposed and implemented in software, and are available in all sorts of languages, including those used for statistical processing and machine learning, such as R and Python. Microsoft Excel, a piece of software that is widely available and used, also has strong regression capabilities.

By now it should be clear that efficient implementations of good optimization algorithms undergird any regression software that is available for use. We have seen how writing an objective function, setting partial derivatives to zero and solving a system of linear equations works for simple cases. To perform optimization in complex cases, more sophisticated algorithms are necessary. These include various flavors of the QR algorithm, gradient descent algorithm including stochastic gradient descent and various optimized versions of it, and coordinate and semi-definite optimization. We will discuss only (stochastic) gradient descent, since a good understanding of such algorithms is necessary to perform research and understand research papers that disseminate research findings. For more sophisticated discussion of regression and optimization, formulating the problem in terms of vectors and matrices is necessary. We do not delve further into matrix formulation in this book.

As we have seen already, given a dataset, it is possible to obtain many different equations that fit, even if they are all linear. We have seen that loss-based regressors produce a family of solutions, and one has to pick a solution that works best. Thus, it is necessary to have a good understanding of measures of goodness and statistical significance so that we can choose a model that we want to use. We discuss only F-test at length in this chapter.

The programs presented in this chapter have all been in R. We have not discussed fundamentals of R in this book and encourage the reader to become familiar with the basics of this language and the various libraries it provides. We plan to provide a website with all the code discussed in this book. Although we have not discussed Python code in this chapter, a student of machine learning is encouraged to learn it since it is the language of choice for many in the machine learning world.

Exercises

1. When various numerical attributes of examples in a dataset are of widely varying ranges, it may be helpful to convert all values to a standard range of some kind. This is called feature scaling. Is feature scaling really helpful in regression? If so, under what conditions? What are some common feature scaling approaches?

2. In this chapter, we obtained a solution for the linear regression problem using a traditional analytical method. However, it is possible to perform linear regression using a greedy algorithm called *gradient descent*. Read up on gradient descent, and mathematically formulate linear regression as a problem that be solved using gradient descent.

3. Set up Ridge, Lasso and Elastic Net regression as problems that can be solved by gradient descent.

4. Some simple non-linear functions can be transformed to linear functions so that linear regression can be performed with them. How can you linearize the following functions? a) exponential equation: $y = \alpha_1 \, e^{\beta_1 \, x}$, b) power equation: $y = \alpha_2 \, x^{\beta_2}$, and c) saturation growth equation: $y = \alpha_3 \frac{x}{\beta_3 + x}$.

5. Given a dataset for regression, we can fit a linear function, a quadratic function or functions of higher degrees. Often, linear regression is preferred over higher order regression. Why is it so? What are pros and cons of linear and higher degree regressions?

6. In regression, multicollinearity refers to dependent variables that are correlated with other dependent variables. Multicollinearity occurs when a model includes multiple dependent variables that are correlated not just to the dependent variable, but also to each other. In other words, it results when you have independent variables that are a bit redundant. How can you identify collinearity in regression? Does the presence of collinearity matter at all while performing regression? If so, What can you do it remove collinearity?

7. When we perform regression, as seen in this chapter, different approaches are likely to result in different equations for regression fit with different variables and coefficients. How can you pick the equation that fits the "best"?

8. In a dataset for regression, outliers may be present. Outliers are points that are extreme values, and they may affect regression results substantially and negatively. How can outliers be found and dealt with during regression?

9. In regression as we have discussed in this chapter, the goal is to find a single continuous function that best fits the data points, and is able to perform prediction for the value of a dependent variable in a generalizable manner. In some cases, instead of one continuous function, it is preferable to fit a number of distinct continuous functions (in small ranges that fit more snugly to the data) that bend in like a single continuous curve. The individual curves are called splines. Discuss how linear, quadratic and cubic splines can be created, given a small dataset. Compare the pluses and minuses of drawing splines compared to a single fitted curve.

10. Regression fits a function (linear or non-linear) that has least cumulative error over all points in a dataset. In contrast, interpolation fits a curve that goes through all points in a dataset, without any error. What are pros and cons of regression versus interpolation? When is one more appropriate compared to the other?

Chapter 3

Tree-Based Classification and Regression

3.1 Introduction

As we know by now, machine learning is a discipline within computer science that attempts to find patterns in large amounts of data. Machine learning always begins with a collection of data of interest to explore. It applies a variety of algorithms to discover hidden regularities or patterns in the collection in the form of a model, and uses the learned model for purposes at hand. A common use is predictive, meaning that the learned model is used to predict the classification or behavior of previously unseen examples. For example, regression, which we discussed in detail in Chapter 2, fits a linear or non-linear function for a dependent variable of the independent variables, and then performs prediction for the value of the dependent variable for a previously unseen data example. Other machine learning models may also infer associations among data examples, or create groups or clusters of tightly related data examples for further analysis.

There are essentially two types of machine learning algorithms: supervised and unsupervised. Supervised algorithms use a collection of data where each example has been examined by a human expert in some ways. Regression is a supervised approach. In regression, the dependent variable is *numeric*. The values of the dependent variable for training examples are obtained from experimental or similar observations. Many supervised algorithms also perform what is called *classification*. The idea of *class* is fundamental to classification. We think of a class as a named group of objects that share some properties or *attributes* or *features*. As we know by now, in machine learning, each example is described in terms of a number of features. For example, if our examples are from the domain of vehicles, each example may have features such as height, length, width, weight, color, runs on gas or diesel, number of seats, etc. The classes may have names such as *hatchback*, *sedan*, *SUV*, *pickup-truck*, *bus*, etc. If our dataset is about classifying loans to individuals as high risk or low risk, the individuals may be described in terms of features such as age, male, female, education level, yearly income and credit rating. In this case, the classes may be called *low-risk* and *high-risk*. Often, it is difficult to define

DOI: 10.1201/9781003002611-3

TABLE 3.1:   A few datasets from the UCI Machine Learning
Repository. The class names are italicized under Description.

Dataset	Description	#Classes	#Features	#Examples
Covertype	Predict forest cover type, e.g., *pine* and *aspen*	7	54	581,012
Glass Identification	Predict the type of glass such as *window, tableware, headlamps*	7	10	214
Iris	Predict the type of iris as one of *setosa, versicolor* and *virginica*	3	4	150
Lenses	Whether a patient should be fitted with contact lenses (*soft* or *hard*) or *not*	3	4	24
Seeds	Whether a kernel is one of three types of wheat: *Kama, Rosa* and *Canadian*	3	7	210

what feature or feature values an object should have to belong to a certain class.

For supervised learning, we need a dataset of "historical" examples, where each example is described in terms of its features and also, the class it is assigned to. Table 3.1 presents a few simple but real datasets that have been used in the machine learning literature. We provide detailed information about the first dataset in Table 3.2. The details have been culled from the UCI Machine Repository site. In this table, we see that there are twelve attributes that are used to describe an example. The first ten are numeric or quantitative, whereas the last two are non-numeric or qualitative, in that they take a small number of discrete values. These are also called categorical attributes. Both these attributes are binary in this case, taking a value of 0 or 1. When we perform machine learning, whether supervised or unsupervised, we should be able to handle both numeric and non-numeric values. The last row shows the class label called *Cover_type*. It takes seven possible values as given in the table.

Given a dataset where each example is described in terms of features and a class label, the task of supervised learning is to discover patterns to build a model that can tell apart each of the classes involved. The model may not be explicit, but stored implicitly. Each machine learning algorithm has a learning bias, and the model formed depends on the learning bias. The learning bias

TABLE 3.2: Attributes in the Covertype dataset from UCI Machine Learning Repository. There are seven class labels in this dataset called cover-type, taking values from 1 to 7.

Description: Predicting forest cover type from cartographic variables. The actual forest cover type for a given observation (30 x 30 meter cell) was determined from US Forest Service data. Independent variables were derived from US Geological Survey and USFS data. The classes are spruce or fir (type 1), lodgepole pine (type 2), Ponderosa pine (type 3), cottonwood or willow (type 4, aspen (type 5), Douglas-fir (type 6) and Krummholz (type 7).

Attribute Information:

Name	Data Type	Measurement
Elevation	quantitative	meters
Aspect	quantitative	azimuth
Slope	quantitative	degrees
Horizontal_Distance_To_Hydrology	quantitative	meters
Vertical_Distance_To_Hydrology	quantitative	meters
Horizontal_Distance_To_Roadways	quantitative	meters
Hillshade_9am	quantitative	0 to 255
Hillshade_Noon	quantitative	0 to 255
Hillshade_3pm	quantitative	0 to 255
Horizontal_Distance_To_Fire_Points	quantitative	meters
Wilderness_Area (4 binary columns)	qualitative	0 or 1
Soil_Type (40 binary columns)	qualitative	0 or 1
Cover_Type (7 types)	integer	1 to 7

refers to the kind of model the algorithm is designed to learn, e.g., straight line separators among classes.

Unsupervised learning also attempts to discover patterns in a dataset. Datasets used in unsupervised learning have examples like datasets used in supervised learning, where examples are described in terms of features. However, examples used in unsupervised learning lack the class label. A common form of unsupervised learning, called *clustering*, attempts to organize examples in various groups, where examples inside a group are similar to each other as much as possible, and dissimilar to examples in other groups as much as possible. To be able to perform such natural grouping, we need to use what is called a *similarity measure* or *similarity metric* to compute how similar or dissimilar two examples are. On the flip side, we can use a *distance measure* or *distance metric* also. It may also be useful to measure the similarity or dissimilarity between two groups of examples.

Researchers and practitioners have come up with a large number of algorithms for supervised and unsupervised learning algorithms. In this chapter and the one that follows, we discuss *classification*, a common form of supervised learning, and Chapter 6 in this book focuses on unsupervised learning.

Among the most commonly used supervised learning or classification algorithms are the nearest neighbor algorithms, tree-based algorithms, support vector machines, probabilistic algorithms based on Bayes theorem, and neural networks. In particular, we discuss tree-based algorithms in this chapter. Beyond the basic tree-based algorithms such as decision trees, it is common these days to use several classifiers that complement each other to perform a number of separate classifications on some datasets, and then use a group decision in terms of voting or consensus, to make the final classification. Such classifiers are called *ensemble* classifiers. They are discussed in this chapter as well, in the context of tree-based classification.

Developing a classifier algorithm, training it on examples, and using it on unseen examples (often, called *testing*) are the main tasks we do in supervised learning. The purpose of using a supervised learning algorithm is to be able to train it on a dataset and use the trained model to classify unseen examples in a test setting or in the real world. To decide which algorithm is best suited for a certain supervised machine learning task, we need to evaluate the performance of the algorithm in terms of metrics. We discuss the most common evaluation metrics for classifiers in this chapter.

3.2   Inductive Bias

A classification algorithm learns a model of the data from the training instances and uses this model to classify previously unseen examples. This model may be an explicit mathematical function of the attributes or more often implicit. Since the number of possible functions to learn is infinite, depending on the algorithm, it often limits itself to a particular class of functions that it seeks to learn from the training examples. This is called the *Inductive Bias*, or simply the *bias* of an algorithm. Thus, each machine learning algorithm usually has some bias to start with. For example, it is possible that a certain classifier is biased from the beginning to draw a straight line or a plane or a hyperplane in a high dimension to separate examples of one class from another. We see such a scenario in 3.1 (a). Here, the small squares represent examples from one class, and the small circles represent examples of a second class. In mathematical terms, the function that such a classifier can learn is simply a straight line or a plane or a hyperplane in higher dimensions. It is a linear classifier. Each training example has two features, simply called x and y here. For example, x can be income in dollars and y can be savings in dollars.

A second example of inductive bias may be that the learner can learn to draw two separating lines that are parallel to the X- and Y-axes. A pair of such lines is used to separate a class's examples from those of another class. Figure 3.1 (b) depicts such a situation. Such a classifier will have to learn the co-ordinates of the point where the two lines meet, with the understanding

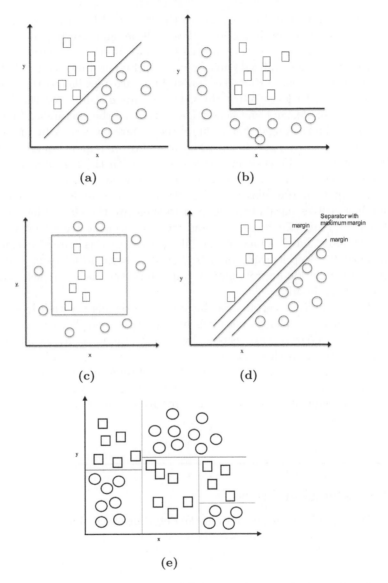

FIGURE 3.1: Types of inductive bias. (a) A classifier may be biased to draw a straight line separating examples of the two classes. (b) The inductive bias may draw two lines that start at a point, are parallel to the two axes, and are directed. (c) The inductive bias may be that the classifier can draw rectangles around the space required by examples of a class. (d) The inductive bias is to draw a separator line or hyperplane like in (a), but with maximum separation between the two classes. (e) The inductive bias is to draw a number of lines parallel to the axes to separate the classes into disjoint rectangular areas. All examples are in 2-D.

(bias) that directed lines parallel to the axes will be drawn for separation of the classes. In higher dimensions, these will be planes or hyper-planes. A third type of bias associated with a classifier may be that it learns a rectangular area (cubes or generalized cubic volumes in higher dimensions) that separates the two classes such that the border lines are parallel to the axes, Such a classifier has to learn four values, the x and y values for the lower corner of the rectangle, and the height and width of the rectangle. Figure 3.1 (c) shows such a rectangle that includes the square classes. A further example of inductive bias we provide is to find a separator region bounded by margins (two- or higher dimensional) on two sides, with the actual separator line or hyperplane exactly in the middle of the region. Figure 3.1 (d) show such a maximum margin classifier. The final example in Figure 3.1 (e) draws a number of lines parallel to the axes to separate the classes into a number of disjoint rectangular bounded areas. In fact, decision trees that we discuss in this chapter, have inductive bias like shown in this figure. Artificial neural networks or deep learning models that we discuss in Chapter 4 are so-called universal function approximations, and thus, the biases of such classifiers are not easy to describe.

It is clear that certain classification algorithms are tightly biased in the sense that the model they learn is highly prescribed. For example, for a classifier that separates two classes by a line (plane or hyper-plane in higher dimensions) has higher bias compared to a decision tree which allows for arbitrary rectangular (cubic or hyper-cubic) volumes, possibly spread throughout the space. To fit a highly biased model, the algorithm may have to accept a higher amount of error compared to fitting a loosely biased model.

3.3   Decision Trees

Often, as humans, when we make decisions regarding classification, we use a tree-like structure to do so. The nodes of such a tree have queries or questions, and the branches lead to different paths based on answers to the queries. We start from the top node, which poses a query, and based on the answer, we take a path down the tree, and we answer another query at the second node, and take another path down the tree based on the answer. We traverse the tree in such a manner to a leaf node where we know which class the example belongs to. As humans, we may be able to build a decision tree ourselves based on the knowledge we possess about the classification problem at hand. However, these days, it is more common to automatically build decision trees based on data in an empirical fashion.

Decision trees are built using supervised learning. In other words, we build a decision tree from a number of labeled examples given to us. Tables 3.4 and 3.3 show the *Lenses* dataset from the UCI Repository. In this section, we

TABLE 3.3: Description of Lenses Dataset from the UCI Machine
Learning Repository

Number of Instances:	24
Number of Attributes:	4 (all nominal or categorical)
Number of classes:	3
	1 : the patient should be fitted with hard contact lenses,
	2 : the patient should be fitted with soft contact lenses,
	3 : the patient should not be fitted with contact lenses.
Attribute 1	age of the patient: (1) young, (2) pre-presbyopic, (3) presbyopic
Attribute 2	spectacle prescription: (1) myope, (2) hypermetrope
Attribute 3	astigmatic: (1) no, (2) yes
Attribute 4	tear production rate: (1) reduced, (2) normal
Class Distribution:	
	1. hard contact lenses: 4
	2. soft contact lenses: 5
	3. no contact lenses: 15

describe how a decision tree can be built from this and other datasets and used
after it has been constructed. Many different decision trees can be constructed
from this dataset. Figure 3.8, later in this chapter, shows a decision tree that
has been built from this dataset.

The Lenses dataset, like any other dataset, is given in terms of features
or attributes for the examples. Each example in the table has five features.
The first four are descriptive, giving us details of an example, whereas the last
one is the class label giving us the class the example belongs to. This *label* is
assigned to an example by an "expert" or "supervisor", and is separated from
the descriptive features by a vertical line in the table. As usual, the descriptive
features may take only categorical values, where the number of distinct values
is small; or, numeric values, where the number of possible values is usually
infinite. The class label usually takes a small number of values. In this case,
the classes are numbered 1, 2 and 3. This and a few other simple datasets are
used in this chapter to illustrate decision tree construction.

3.3.1   General Tree Building Procedure

The task of building a decision tree involves deciding which one among
the descriptive features should be used to construct the first query, situated
at the root node. The query will involve a feature or attribute and a value
for the feature, along with a comparison operator as seen in Figure 3.3. The
first question we need to ask is which feature among the ones available should

TABLE 3.4: Lenses Dataset from the UCI Machine Learning Repository

No	Att1	Att 2	Att 3	Att 4	Class
1	1	1	1	1	3
2	1	1	1	2	2
3	1	1	2	1	3
4	1	1	2	2	1
5	1	2	1	1	3
6	1	2	1	2	2
7	1	2	2	1	3
8	1	2	2	2	1
9	2	1	1	1	3
10	2	1	1	2	2
11	2	1	2	1	3
12	2	1	2	2	1
13	2	2	1	1	3
14	2	2	1	2	2
15	2	2	2	1	3
16	2	2	2	2	3
17	3	1	1	1	3
18	3	1	1	2	3
19	3	1	2	1	3
20	3	1	2	2	1
21	3	2	1	1	3
22	3	2	1	2	2
23	3	2	2	1	3
24	3	2	2	2	3

be the feature for the question, i.e., on what basis should it be selected? The next question is what value of the feature is relevant to ask the question and what the comparison operator should be. The comparison operator is usually equality for text or a categorical feature, and $>$, $<$ or $\geq$, or $\leq$ for a numerical feature. It is usually not a good idea to use exact equality for numeric features.

The objective is to build a compact decision tree that is consistent with all the examples from which the decision tree is built. The process is called *training* the decision tree. If we want to build the most compact tree, the procedure becomes an exponential time process since the total number of possible trees we can build from n examples, each with k features with each feature taking a number of values is exponential in number. Thus, it is not judicious to create all possible trees and choose the most compact or optimal tree. Therefore, the approach to build a decision tree is greedy, deciding on a feature to use in a query quickly that cannot be undone once a decision has been made. Decisions regarding splitting features and values are made sequentially. Each is greedy and decisions made later depend on prior decisions that have been committed. Thus, it is quite likely that a decision tree built by one of the commonly used methods is only locally optimal and is not the global optimal one. However, even such locally optimal trees happen to be reasonably good classifiers.

3.3.2 Splitting a Node

Every node of the tree is associated with a data subset, with the root node associated with the entire dataset. The choice of a feature or attribute to construct a query at a certain node of the tree is usually made using a metric that can tell us how the data associated with the node are dispersed, i.e., how homogeneous or non-homogeneous that data are, based on the class labels. The concept of homogeneity is discussed below. Two metrics are commonly used: the Gini index and Entropy. We discuss the use of both in this chapter.

3.3.2.1 Gini Index

The Gini index has been traditionally used to compute a measure of economic inequality that exists in a society or country. The Gini index has been adapted for use in the context of a dataset where we have discrete labels or classes for the examples. Suppose we have a dataset with N examples such that the examples belong to K distinct classes, $c_1 \cdots c_K$. The Gini index for a dataset D containing N examples, with n_i examples from class c_i, $i = 1 \cdots K$, is

$$Gini(D) = 1 - \sum_{i=1}^{K} \hat{p}_i^2 \qquad (3.1)$$

where $\hat{p}_i$ is the empirical estimate of probability for class c_i. It is empirical in the sense that it is computed from the training dataset (and, not general knowledge about the entire population) such that $\hat{p}_i = \frac{n_i}{N}$ where n_i is the number of examples of class c_i in the dataset that has a total of N examples. If the dataset has examples from only two classes c_1 and c_2,

$$Gini(D) = 1 - \hat{p}_1^2 - \hat{p}_2^2 \qquad (3.2)$$

where $\hat{p}_1 = \frac{n_1}{N}$ and $\hat{p}_2 = \frac{n_2}{N}$, with $n_1 + n_2 = N$. In the case of two classes, we can perform some algebra and obtain a formula for $Gini(D)$ that we can plot.

$$
\begin{aligned}
Gini(D) &= 1 - \hat{p}_1^2 - \hat{p}_2^2 \\
&= 1 - \hat{p}_1^2 - (1 - \hat{p}_1)^2 \\
&= 1 - \hat{p}_1^2 - (1 + \hat{p}_1^2 - 2\hat{p}_1) \\
&= 2(\hat{p}_1 - \hat{p}_1^2) \qquad (3.3)
\end{aligned}
$$

We can plot $Gini(D)$ with one of the probabilities $\hat{p}_1$ on the X-axis with the understanding that $\hat{p}_2 = 1 - \hat{p}_1$. How this function varies as the probabilities change is shown in Figure 3.2. Thus, as the probability $\hat{p}_1$ of the first class

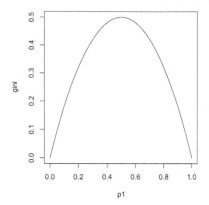

FIGURE 3.2: Gini Graph. The X-axis shows the fraction of the dataset in a class where there are two classes. The Y-axis shows the Gini index of the dataset.

increases starting from approximately 0 when $\hat{p}_1 = 0$, the value of the Gini index increases, reaching approximately 0.5 when $\hat{p}_1 = 0.5$, and then falling back to approximately 0, when $\hat{p}_1 = 1$. We know that there are two classes and hence the probabilities for the two classes add up to 1. When the dataset D is balanced between the two classes, i.e., $\hat{p}_1 \approx \hat{p}_2 \approx 0.5$, the Gini index is the highest. In other words, balanced datasets have high Gini score. When the dataset is unbalanced in the distribution of the two classes, e.g., when $\hat{p}_1$ is low and $\hat{p}_2$ is high, or vice versa, the value of Gini index is low.

When we build a decision tree, we are given a dataset where the examples of the classes are mixed. The objective is to build a tree where at the root node and each of the interior nodes, there are questions that lead to a split of the associated data subset, ultimately leading to leaf nodes that are homogeneous. In other words, each leaf node has a homogeneous subset of data associated with it, i.e., all examples in the data subset associated with a leaf node belong to only one of the classes.

We build the decision tree recursively. Thus, the procedure followed at the root node is repeated at all interior nodes as the tree is built. The entire dataset is associated with the root node. We start building the tree by creating branches at the root node. In other words, we split the entire dataset into subsets based on a question we ask with respect to one of the features and the values of the chosen feature. The dataset usually has several or many features. The question to ask at the root node is which one of the features should be used considering the entire dataset is associated with the node, and also what value for this feature should be used. For now, we have assumed that each feature takes one of a few possible discrete values, and there will be a branch for each value. The Gini index is used to decide in a greedy manner which one of the k features is used to split at the root node. Thus, at the root node, we look at all the features and for each feature look at all of its possible values and perform exploratory splits to see which one is the best. Our goal is to build a compact decision tree, and hence at the root node, our sub-goal is to produce the split that changes, most downwardly, the Gini index associated with the subset of data at the node and the proposed splits of this subset. This is explained with an example below.

To select a feature to create a question, we compute how the Gini index changes when we use various features to separate out the data subset into further subsets.

Non-numeric or Discrete-valued Features: A discrete-valued feature takes a small number of enumerated values. For example, in the Lenses dataset, each of the features takes one of a discrete number of values. The age of a patient is not given as a number, but in terms of three enumerated values, *young*, *pre-presbyopic*, and *presbyopic* (when the eyes lose the ability to see things clearly up close). These can be equivalently written as discrete integers: 1, 2 and 3, respectively if we so want, with 1 standing for pre-presbyopic, 2 for pre-presbyopic and 3 for presbyopic. Similarly, the feature called astigmatism takes two values *no* and *yes*, which can also be written as 1 and 2, respectively, if we desire. It is actually a binary feature and could have been written as 0 and 1 as well. But, what is important to know is that we cannot write a value for astigmatism as a value other than a small number of prescribed values; so nothing like 0.1 or 2.1 is allowed.

If we use a discrete-valued feature f_j to separate the data subset D into further disjoint subsets, we can compute the Gini index of the entire subset D before split, and then after split—the weighted sum of indices of the further subsets that result, and then the change in Gini index value when the feature f_j is used for the process, using the formula below:

$$\Delta Gini(D, f_j) = Gini(D) - \sum_{i=1}^{nv(f_j)} \frac{|D_{f_i}|}{|D|} Gini(D_{f_i}) \qquad (3.4)$$

where D is the original dataset, $nv(f_j)$ is the number of distinct values feature f_j takes, $|D|$ is the number of examples in dataset D, and $|D_{fi}|$ is the number of examples in the ith branch of the tree, i.e., the number of examples in D that have the ith unique value of f. If f_j can take m_j possible values, we will have m_j branches. As a specific example, if we choose Attribute 1 (age of patient) as the feature to use as a question, we have three branches as show in Figure 3.3 (b) because this feature has three possible values. If we choose Attribute 2 (spectacle prescription) as the question feature, we have two branches as shown in Figure 3.3 (c), because the attribute has two possible values.

If we use Attribute 1 as the feature at the top node, we have three branches. At the top node, we have the entire Lenses dataset associated with the node. We have a subset associated with each of the branches, the specific subset

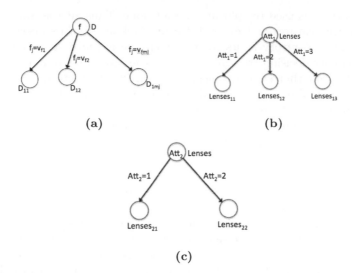

(a)   **(b)**

(c)

FIGURE 3.3: (a) Branching dataset D on feature f_j that has m_j possible values. (b) Branching dataset *Lenses* on feature Attribute 1 that has 3 possible values. (b) Branching dataset *Lenses* on feature Attribute 2 that has 2 possible values.

depending on the value of Attribute 1. Let us call the subset associated with the value 1 of Attribute 1 be called $Lenses_{11}$, the subset associated with value 2 of feature be called $Lenses_{12}$, and the subset associated with value 3 of the feature be called $Lenses_{13}$. We can compute the Gini index of each of these subsets, and then the change in Gini index if we divide up the Lenses dataset with Attribute 1 as follows. When we compute the Gini index for the group of resultant subsets, we weigh each individual subset's index by the proportion of examples in the subset, considering the entire set.

$$Gini(Lenses) \quad = 1 - \left(\frac{4}{24}\right)^2 - \left(\frac{5}{24}\right)^2 - \left(\frac{15}{24}\right)^2 = 0.53819$$

$$Gini(Lenses_{11}) \quad = 1 - \left(\frac{2}{8}\right)^2 - \left(\frac{2}{8}\right)^2 - \left(\frac{4}{8}\right)^2 = 0.625$$

$$Gini(Lenses_{12}) \quad = 1 - \left(\frac{1}{8}\right)^2 - \left(\frac{2}{8}\right)^2 - \left(\frac{5}{8}\right)^2 = 0.53125$$

$$Gini(Lenses_{13}) \quad = 1 - \left(\frac{1}{8}\right)^2 - \left(\frac{1}{8}\right)^2 - \left(\frac{6}{8}\right)^2 = 0.40624$$

Now, we can compute change in Gini index when we divide the dataset into three subsets using Attribute 1 as follows.

$$\Delta Gini(D, f_1) = Gini(Lenses) - \frac{4}{24} Gini(Lenses_{11}) - \frac{5}{24} Gini(Lenses_{12})$$
$$- \frac{15}{24} Gini(Lenses_{13})$$
$$= 0.53819 - \frac{1}{24}(4 \times 0.625 + 5 \times 0.53125 + 15 \times 0.40624)$$
$$= 0.06944$$

We can compute the change in Gini index for all the features for all their possible discrete values, and choose the one that produces the highest amount of Gini change. Thus, if we have k features in the examples in the dataset and the features on average take $\overline{m}$ possible discrete values, we have to compute the change in Gini index for each $\overline{m}k$ possible feature-value combinations to compute Gini indices for the subsets that result. In other words, we choose the feature that separates the mixed dataset into the least mixed group of subsets of data among all split subsets possible. This is because our ultimate objective in building a decision tree is to obtain a tree where all the leaves have homogeneous subsets of data, i.e., each leaf node is associated with a number of examples corresponding to only one of the possible classes. This is one step toward the goal of building an "optimal" decision tree. Decisions at individual nodes are made independently of other nodes, and thus, there is no global concerted decision making. And, as noted earlier, once a decision to split has been made, there is no way to undo it.

Note that when computing Gini gain, we subtract the weighted Gini index value for the resulting split subsets from the Gini index value for the original unsplit set. Since the subtraction is done for each case before comparison, it is not necessary to do the subtraction to compare. We can simply compare the weighted Gini index values of the resulting subsets and pick the subset that has the smallest Gini value, i.e., the split that results in the most homogeneous or more pure split among all the splits.

Numeric Features: In the Lenses dataset, each feature takes a small number of discrete values. But, feature values can be non-discrete or numeric as well. Consider the Iris and Diabetes datasets from the UCI Machine Learning Repository. The Iris dataset has two numeric features: petal length, and petal width for Iris flowers. Each example is labeled to belong to one of three classes of iris flowers: *setosa*, *versicolor* or *virginica*. A few examples of the dataset are given in Table 3.5. The actual dataset has 150 rows.

When we handle a numeric attribute while building a decision tree, the decision to split can be made using the Gini index, but the candidates for split need to be obtained in a different way. The questions to be asked at a node can be of the form $x < x_{split}$, $x \leq x_{split}$, $x > x_{split}$ or $x \geq x_{split}$, where x_{split} is the partitioning value. The split is binary. First, the values of the feature x need to be sorted in ascending order. Let the range of values of x be the

TABLE 3.5:   Iris Dataset from the UCI Machine
Learning Repository. sLength is sepal length, sWidth
is sepal width, pLength is petal width, pType is petal
width, and iType is the type of iris.

sLength	sWidth	pLength	pWidth	iType
5.1	3.5	1.4	0.2	setosa
4.9	3.0	1.4	0.2	setosa
4.7	3.2	1.3,	0.2	setosa
4.6	3.1	1.5	0.2	setosa
5.0	3.6	1.4	0.2	setosa
5.4	3.9	1.7	0.4	setosa
4.6	3.4	1.4	0.3	setosa
5.0	3.4	1.5	0.2	setosa
4.4	2.9	1.4	0.2	setosa
4.9	3.1	1.5	0.1	setosa

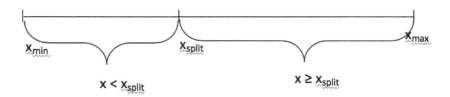

FIGURE 3.4: Splitting a numeric-valued feature.

closed interval $[x_{min}, x_{max}]$, meaning x_{min} and x_{max} are part of the interval.
Any value in the interval can be potentially proposed as x_{split}, the point at
which the question is asked, as shown in Figure 3.4. How x_{split} is chosen is
heuristic, since it is not generally possible to look at every possible value in
the range as a potential split point. If the number of points is small, one could
potentially treat every value as a potential split point. However, especially
if the dataset is large, we can divide the points into quartiles (where, each
quartile has about a quarter of the values), or deciles (where, each deciles has
about a tenth of the values), etc. It is also possible to perform binary search
up to a certain depth to compute the split point. How it is done depends on
the implementation. Thus, a number of candidate split points are chosen, and
at each point, the gain in Gini index is computed.

   If we have a number of numeric features only, for each proposed split point
for each of the features, the gain in Gini index is computed as before. Each
split for a numeric feature is binary. If we have k numeric features and for
each feature's values, $\overline{m}$ splits are considered on average, we need to compute
Gini index changes for $\overline{m}k$ cases. The <feature, split point> combination that

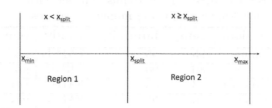

FIGURE 3.5: Regions formed by a split at $x = x_{split}$.

gives the best Gini gain, considering all features and their potential splits, is used to perform the actual split.

As shown in Figure 3.5, each question asked at a node has a numeric feature on the left, an inequality, and a numeric value on the right-hand side. This means that if we think of the feature value changing along an axis, a vertical line separates the two regions into which the value separates the entire space. Assuming we have only one feature x, Figure 3.5 shows a split on the value of a feature x at x_{split} into two regions by a vertical line at x_{split}, which we call Region 1 and Region 2. Region1 may be divided into further regions by a line that is perpendicular to the x-axis by questions we may ask later. Region 2 may also be divided into additional regions by other lines perpendicular to the x-axis in the future. Such divisions could potentially continue recursively till the space has been divided into regions that are homogeneous. If we have two features, the separated regions will have bounding lines that are parallel to either of the two axes. If we have three or more features, the separation into regions will be done by planes or hyper-planes parallel to the axes. Whether the equality is on the left or the right partition depends on the user's decision, based on knowledge of the domain.

A tree has a number of questions at the nodes, and all the questions taken together divide the space of feature values examples can take into a number of spaces with lines or (hyper-)planes that are parallel to the axes. This is the classification bias of decision tree algorithms.

Mixed Numeric and Non-Numeric Features: If we have a mix of discrete and numerical features, for each discrete feature, the potential splits are of the type $x = x_{value}$ whereas for numerical features, the potential splits are of the form $x < x_{value}$ (assuming we using the $<$ relation in the split question). All the potential splits are compared to pick the one that gives the biggest Gini gain.

A simple dataset with both numeric and non-numeric features is shown in Table 3.6. It is a small dataset, first used by [20]. The dataset is available in several incarnations, one with all non-numeric features, and the one here with a mixed set of features. The dataset makes observations about weather conditions and whether one plays tennis in such situations. The goal is to learn and predict whether one should play tennis in situations that have not

TABLE 3.6:   Play Tennis Dataset with
Mixed Numeric and Non-numeric Features

outlook	temp	humidity	windy	play
sunny	85	85	FALSE	no
sunny	80	90	TRUE	no
overcast	83	86	FALSE	yes
rainy	70	96	FALSE	yes
rainy	68	80	FALSE	yes
rainy	65	70	TRUE	no
overcast	64	65	TRUE	yes
sunny	72	95	FALSE	no
sunny	69	70	FALSE	yes
rainy	75	80	FALSE	yes
sunny	75	70	TRUE,	yes
overcast	72	90	TRUE	yes
overcast	81	75	FALSE	yes
rainy	71	91	TRUE	no

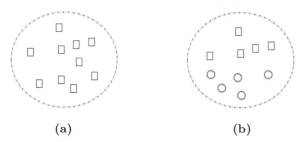

(a)   (b)

FIGURE 3.6: (a) A dataset with all examples from the same class, with a dataset entropy of 0. (b) A dataset with half the examples from one class and the other half from another, with an entropy of 1.

been encountered before. In this dataset, the feature called *outlook* takes three values: *sunny, overcast* and *rainy*; and the feature called *windy* takes two values: *TRUE* and *FALSE*. The features called *temp* and *humidity* take numeric values. The class label is called *play* and takes values *yes* and *no*.

3.3.2.2   Entropy

So far, we have discussed how to split nodes using the Gini index. Another metric that has been used frequently to split tree nodes is *entropy*. Entropy also measures the amount of chaos or dissimilarity we have in a dataset, or how mixed up or inhomogeneous a dataset is. As before, homogeneity is measured with respect to the class(es) the examples in the dataset belong to. For example, if we have a dataset of N examples, and all examples belong to the same class c, we will say that the entropy of the dataset is 0 (See Figure 3.6

(a)). In this case, we have 10 examples in the dataset included inside the big dashed circle, and each data example is a small rectangle, and thus the class is homogeneous. If we have a dataset of N examples, and $\frac{N}{2}$ examples belong to a class c_1 and $\frac{N}{2}$ examples belong to a class c_2, we can say that the dataset is mixed up to the highest degree possible and its entropy is 1. See Figure 3.6 (b)) where the dataset of 10 contains 5 examples of the circle class and 5 from the rectangle class. Entropy takes the highest value when the dataset is all mixed up, and the least value when the dataset is homogeneous, just like Gini index. Thus, the two metrics are quite similar in nature.

The entropy of a dataset D containing n examples, with n_1 examples from class c_1 and n_2 examples from class c_2 is

$$Entropy(D) = -\frac{n_1}{N} \ log\frac{n_1}{N} - \frac{n_2}{N} \ log\frac{n_2}{N} \tag{3.5}$$

$$= -\hat{p}_1 \ log \ \hat{p}_1 - \hat{p}_2 \ log \ \hat{p}_2 \tag{3.6}$$

$$= \sum_{i=1}^{2} -\hat{p}_i \ log \ \hat{p}_i \tag{3.7}$$

where $n_1 + n_2 = N$ and $\hat{p}_1 = \frac{n_1}{N}$ and $\hat{p}_2 = \frac{n_2}{N}$. $\hat{p}_1$ is the probability of a data example being in class c_1 based on the observed dataset, and $\hat{p}_2$ is the probability of a data example being in class c_2. If the dataset contains examples from k classes $\mathcal{C} = \{c_1 \cdots c_k\}$, the entropy of the dataset is

$$Entropy(D) = \sum_{i=1}^{k} -\hat{p}_i \ log \ \hat{p}_i \tag{3.8}$$

where $\hat{p}_i = \frac{n_i}{N}$ with n_i being the number of examples in class c_i in the dataset, with $\sum_{i=1}^{k} \hat{p}_i = 1$ and $\sum_{i=1}^{k} n_i = N$. Thus, the entropy of a dataset considering the classes of examples in it lies between 0 and 1. In contrast, the Gini index, discussed earlier, lies between 0 and 0.5. If a dataset is not as mixed as being equally divided between two (or more classes), the entropy value is less than one. Thus the entropy curve looks like the one given in Figure 3.7, assuming we have two classes to choose from. The X-axis shows the proportion of examples from one of the two classes in the dataset and the Y-axis shows the entropy of the dataset. If we have two classes, the maximum value of entropy is 1 and the minimum value is 0. If there are k classes in the dataset, the maximum value of entropy is $log_2 \ N$ and the minimum value remains 0.

Similar to our discussion with Gini index, to select a feature to create a query, we do not look at entropy of a dataset directly, but compute how the entropy will change when we use the different features to separate out the dataset into subsets. If we use a feature f to separate the dataset into disjoint subsets, we can compute the before separation entropy of the entire set, and the after separation weighted sum of entropies of the subsets, and then the

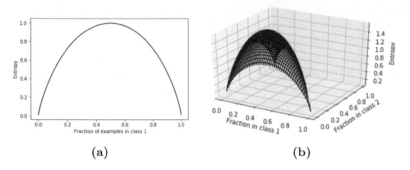

FIGURE 3.7: Entropy Graph. (a) The X-axis shows the fraction of the dataset in a class where there are two classes. The Y-axis shows the entropy of the dataset. (b) The case with three classes.

change in entropy when the feature f_j is used for the process.

$$\Delta Entropy(D, f_j) = Entropy(D) - \sum_{i=1}^{nv(f_j)} \frac{|D_{f_i}|}{|D|} Entropy(D_{f_i}) \qquad (3.9)$$

where D is the original dataset, $nv(f_j)$ is the number of distinct values feature f_j can take, $|D|$ is the number of examples in dataset D, and $|D_{f_i}|$ is the number of examples in the ith branch of the tree, i.e., the number of examples in D that have the ith unique value of f. This formula is similar to Equation 3.4.

Just like we did for Gini index earlier, we can obtain the gain in entropy as we propose splits based on values of the discrete attributes of the Lenses dataset. The entropy gain computations when we split on Attribute 1 (age of patient) are given below.

$$Entropy(Lenses) \quad = -\frac{4}{24} log_2 \frac{4}{24} - \frac{5}{24} log_2 \frac{5}{24} - \frac{15}{24} log_2 \frac{15}{24} = 1.3261$$

$$Entropy(Lenses_{11}) \quad = -\frac{2}{8} log_2 \frac{2}{8} - \frac{2}{8} log_2 \frac{2}{8} - \frac{4}{8} log_2 \frac{4}{8} = 1.5000$$

$$Entropy(Lenses_{12}) \quad = -\frac{1}{8} log_2 \frac{1}{8} - \frac{2}{8} log_2 \frac{2}{8} - \frac{5}{8} log_2 \frac{5}{8} = 1.3844$$

$$Entropy(Lenses_{13}) \quad = -\frac{1}{8} log_2 \frac{1}{8} - \frac{1}{8} log_2 \frac{1}{8} - \frac{6}{8} log_2 \frac{6}{8} = 1.0613$$

Now, we can compute change in entropy when we divide the dataset into three subsets using Attribute 1 as follows. Note that each of the three subsets are of equal size.

$$\Delta Entropy(D, f_1) = 1.3261 - \frac{1}{24}(4 \times 1.5000 + 5 \times 1.3844 + 15 \times 1.0613) = 0.1244$$

We can compute the change in entropy for all the features that an example has, and choose the one that produces the highest amount of entropy change. In other words, we choose the feature that separates the mixed dataset into the least mixed subsets of data. As explained earlier during discussion of splitting using Gini index, this is because our ultimate objective in building a decision tree is to obtain a tree where all the leaves have homogeneous subsets of data, i.e., each leaf node corresponds to a number of examples corresponding to only one of the possible classes.

Once we have selected the first feature to ask question at the top node of the decision tree, we build branches out of this node taking into account the distinct values a categorical feature takes. We associate all the data examples at the top node, and pass along examples associated with the appropriate feature values down the branches. Thus, we will have three children nodes of the top node, each associated with a smaller number of examples. We repeat the process of building a subtree by selecting a head node for the subtree using entropy change computation as described above. We stop building a subtree below a node if all the data examples associated with it become homogeneous i.e., their class becomes the same. If the feature takes a continuous value, the computation of the split point requires a heuristic search as discussed in the case of Gini index.

3.4 Simple Decision Trees in R

There are several libraries in R that can be used to build decision trees. We discuss a few libraries here. We start with the `tree` library that builds decision trees with the Gini index. We show how trees can be built when the features are all discrete, all numeric, and mixed between discrete and numeric. When we discuss building a tree with numeric attributes, we illustrate the inductive bias of decision tree classifiers. There are other popular libraries such as `rpart`, `party` and `ctree` that can also be used to build trees. We will also measure the performance of the tree building process using some R packages.

3.4.1 Simple Decision Trees Using `tree` Library with Gini Index and Discrete Features

We first consider the situation where we have only non-numeric attributes. The `Lenses` dataset we have been discussing is such a dataset. Let us look at how we can build a decision tree using the library `tree`.

```
1  library(tree)
2  contactLenses <- read_csv("Datasets/txt/contactLenses.txt")
3  View(contactLenses)
4
```

```
 5 #R reads the column values as strings
 6 #We need to convert the strings into factors, i.e., enumerated values
 7 contactLenses$age <- as.factor (contactLenses$age)
 8 contactLenses$prscr <- as.factor (contactLenses$prscr)
 9 contactLenses$astg <- as.factor(contactLenses$astg)
10 contactLenses$tearrt <- as.factor (contactLenses$tearrt)
11 contactLenses$class <- as.factor (contactLenses$class)
12
13 #Build all the way to the bottom till every node is homogeneous
14 #minsize is the minimum number of data points at a node
15 lenseTree <- tree (formula = contactLenses$class ~ .,data = contactLenses,
16          split = c("gini"), minsize=1)
17
18 summary (lenseTree)
19 plot (lenseTree)   #plot the tree with no labels
20 #to get the proper labels that come from the dataset
21 text(lenseTree, pretty=0)
```

Line 2 of the program reads the data file and line 3 views the contents. The first four attributes of the examples are read as integers, and the class is read as a character string. For R to work with these as enumerated values, we need to convert them to what are called factors in R. Lines 7-11 do that. Building the tree is quite simple in R. We do so in lines 15 and 16 of the program. R uses the "formula" syntax to build trees, just like formulas we used in building regression fits in Chapter 2. The statement

```
lenseTree <- tree (formula = contactLenses$class ~ .,data = contactLenses,
      split = c("gini"), minsize=1)
```

builds a tree from all the attributes available in the dataset. The dot (.) signifies all attributes. If we had wanted, we could have selected features we want to use and name them on the right side of ~, separated by +. We build a tree that predicts the class attribute, i.e., what kind of glasses to wear. It uses Gini index as the splitting criterion. We use `minsize = 1` to ensure that the tree building continues till every leaf node is homogeneous. If necessary, a leaf node may be associated with a single example necessary. We can build the tree using the `plot` function, which does not label the branches or the nodes, which is done by the `text` command. The tree drawn and labeled by the program above is given in Figure 3.8. It is a tree with 7 leaf nodes. The root asks the question if `tearrt` or tear production rate is equal to 1 or is reduced. If so, it takes the left branch and if not, it takes the right branch. If the tear production is reduced, no contact lenses are necessary. If the tear production rate is not reduced, further questions are needed to make a decision. The next question it asks is if the value of astigmatism is 1, i.e., there is no astigmatism, it follows the left branch, otherwise the right branch. The tree is built in this fashion till every leaf node becomes homogeneous.

The `summary` command in R prints the following.

```
Classification tree:
tree(formula = contactLenses$class ~ ., data = contactLenses,
    split = c("gini"), minsize = 1)
```

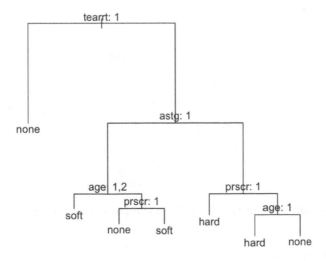

FIGURE 3.8: The full tree built from the Lenses dataset using Gini index.

```
Number of terminal nodes:  7
Residual mean deviance:  0 = 0 / 17
Misclassification error rate: 0 = 0 / 24
```

It shows that a classification tree was built with 7 terminal nodes and with no misclassification. It provides some statistics also. Since the tree has been built down to leaves that may contain single examples, the misclassification rate is 0. In other words, the tree correctly classifies each node in the training set. This is, in fact, self-evident since we have built the tree from the entire training set and have not validated it in any other way. We will see how we can validate trees as they are built later in the chapter.

3.4.2 Simple Decision Trees Using tree Library with Gini Index and Numeric Features

If all the features are numeric, like in the Iris dataset, we can still use the Gini index for splitting. The tree obtained for the Iris dataset is shown in Figure 3.9. In this figure, we see that every question asked is of the form $x < x_{split}$, and unlike a tree built from discrete attributes only where an attribute is used only once in building the tree, in the case of numeric attributes, the same attribute can be used repeatedly, obviously with different values.

The summary of a tree built for the Iris dataset, given by the tree library is given below.

```
Classification tree:
tree(formula = iris$iType ~ ., data = iris, split = c("gini"),
```

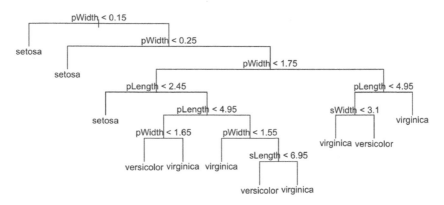

FIGURE 3.9: The full tree built from the Iris dataset using the Gini index.

```
    minsize = 1)
Number of terminal nodes:   11
Residual mean deviance:   0 = 0 / 139
Misclassification error rate: 0 = 0 / 150
```

In Figure 3.9, we see that the features used to build the tree are pWidth and pLength at the top several levels, and toward the bottom, the features sWidth and sLength are used. This tells us that pWidth and pLength are more important for the classification than the other two features. In the tree we built in the R program, we have potentially used all the features in the datasets by using the class variable on the left side of the formula and then just the dot . on the right hand side. The dot signifies all features. It is possible to visualize how a decision tree partitions the space of examples it classifies, particularly so if the attributes are numeric.

Since we have 4 features in the Iris dataset, it is not possible to draw how the space is partitioned by the decision tree built in four dimensions. To visualize how a decision tree partitions space when all the features are numeric, we can use one or two features for illustrative purpose. Below, we use only the two most influential features to build the tree,

```
irisTree <- tree (formula = iris$iType ~ iris$pWidth + iris$pLength, data = iris,
        split = c("gini"), minsize = 1)
```

to obtain the tree given in Figure 3.10. Here the formula asks to build a tree to classify iType using the pWidth and pLength attributes only, for the iris dataset. We can then use the command

```
partition.tree(irisTree)
```

to show the tree partitions the space. The partitions created by this tree are given in Figure 3.11. The partition illustration is in 2-D since we have

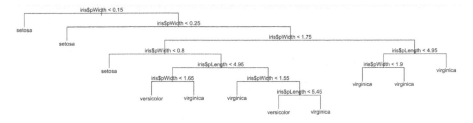

FIGURE 3.10: Iris tree built using two features only.

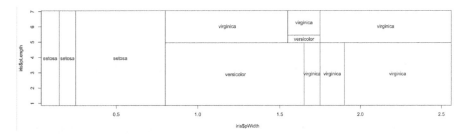

FIGURE 3.11: Split of the space into regions when an Iris tree is built using two features only.

two features represented, with values of `pWidth` along the X-axis and those of `pLength` along the Y-axis. The space of values has been partitioned into a number of rectangular regions with boundaries parallel to the axes. The question at the root of the tree in Figure 3.10 is $pWidth > 0.15$ and if the answer is yes, the class is setosa. The region in the space corresponding to the yes answer is show as the leftmost rectangle in the partition diagram. The no answer corresponds to the region $pWidth \leq 0.15$, and is further sub-divided into homogeneous regions, each region corresponding to a leaf node.

3.4.3 Simple Decision Trees Using `tree` Library with Gini Index and Mixed Features

If we have a mixed set of features, the approach taken is the same. The potential splits of the discrete features are at each of the discrete values. The potential splits of the numeric features are obtained heuristically whereas as discussed before, the potential splits of the non-numeric features are obtained at every discrete value.

The summary of a tree built for the Iris dataset, given by the `tree` library is given below. The actual tree is given in Figure 3.12. Notice that the top level node is branched on the discrete feature `outlook`, with either `rainy` or `sunny` values causing the left branch; the right branch is corresponding to the `overcast` value leads to a `yes` answer. On the second level,

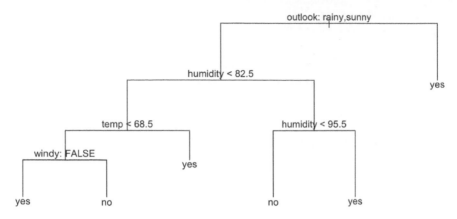

FIGURE 3.12: The full tree built from the Playtennis mixed dataset using the Gini index.

`humidity < 82.5` is a question on a numeric attribute, leading to a left traversal for a yes answer and a right branch for a no answer. The summary returned by the `tree` library in R is given below.

```
Classification tree:
tree(formula = weather_numeric$play ~ ., data = weather_numeric,
    split = c("gini"), minsize = 1)
Number of terminal nodes:  6
Residual mean deviance:  0 = 0 / 8
Misclassification error rate: 0 = 0 / 14
```

3.5   Overfitting of Trees

There are many issues that arise when building a useful decision tree that can be used in practice. The basic algorithm can be modified to address these issues. We consider one particular grave issue in this section.

The problem we discuss is how to handle *overfitting*. The purpose of a machine learning algorithm is to *generalize* from a limited number of examples. To do so, an algorithm should make an attempt to get rid of any accidental similarities that may exist so that the trained algorithm can be used on unseen examples or released to the public to use on whatever dataset they may have. The term *overfitting* means learning too much from the training data as if to memorize (aspects of) the data. Thus, an overfit decision tree is built to "perfection", considering the training data. In other words, an overfit tree is a built tree where every leaf is homogeneous, even if a node is associated with a single data example. Such a tree has possibly learned every quirk that may be present in the training data, in addition to learning the generalizable essence of the relevant concepts.

A machine learning algorithm must generalize by the right amount trying to avoid overfitting, which is difficult to do. It is possible that we build a large decision trees with many levels and many nodes to fit the training data exactly. For example, we have built trees in Figures 3.9 and 3.10 that fit the iris data exactly; they are different trees, but each has 11 nodes. It is quite likely that such a tree is too detailed and it has built nodes to cater to accidental regularities that are present in this dataset only and may not be present in another dataset that we may collect for the same phenomena later or someone else may have collected. Thus, it is possible that when put to action after training, a decision tree with 5 nodes (say) may perform as well as or better than a decision tree with 11 nodes. The smaller tree may perform a little worse on the training data, but may work better on unseen data that it needs to classify. One way to build a decision tree with the right amount of nodes so as not to overfit but learn the generalization involves pruning the decision tree. There are several ways to prune a decision tree, but the preferred way is to build a complete decision tree with as many nodes as necessary to get to homogeneous leaf nodes, but then remove as many nodes as possible with the goal of removing nodes created to cater to accidental regularities. One may simply pick and delete a random node and all nodes below it, and test the new reduced tree using the same dataset that was used to test the entire tree, and if the new reduced tree's performance is not worse than the original tree, keep the reduced tree. The process of pruning is repeated a number of times, and a node is removed only if performance does not degrade on a test dataset. Thus, it may be hypothesized that the nodes eliminated from the original tree covered only non-essential and possibly, accidental regularities and their removal actually not only does not degrade performance, but makes the performance more efficient (since fewer nodes need to be tested for an unseen example) and at the same time possibly a better performer on unseen examples.

3.5.1   Pruning Trees in R

So far, we have been using the `tree` library in R to build our trees. The `tree` library provides us with several ways to perform pruning. A simple way to prune is to provide the number of nodes we want in the tree, once we have built the tree to its fullest. Let us go back to the R statement that builds a complete tree from the Lenses dataset for our examples.

```
tree(formula = contactLenses$class ~ ., data = contactLenses,
    split = c("gini"), minsize = 1)
```

Pruning can be performed using the `best` option to the `prune.tree` command. Let us build trees by pruning down the tree to 2, 3 and 5 nodes and see what happens. The `method` argument says that it should use misclassification rate for pruning. In other words, it attempts to prune to reduce the misclassification rate or maximize the classification accuracy.

```
prunedLenseTree2 <- prune.tree (lenseTree, method = c ("misclass"), best=2)
prunedLenseTree3 <- prune.tree (lenseTree, method = c ("misclass"), best=3)
prunedLenseTree5 <- prune.tree (lenseTree, method = c ("misclass"), best=5)
```

We can then print the summary of these trees as follows.

```
> summary (prunedLenseTree2)

Classification tree:
snip.tree(tree = lenseTree, nodes = 3L)
Variables actually used in tree construction:
[1] "tearrt"
Number of terminal nodes:  2
Residual mean deviance:  1.176 = 25.86 / 22
Misclassification error rate: 0.2917 = 7 / 24
> summary (prunedLenseTree3)

Classification tree:
snip.tree(tree = lenseTree, nodes = 6:7)
Variables actually used in tree construction:
[1] "tearrt" "astg"
Number of terminal nodes:  3
Residual mean deviance:  0.6212 = 13.04 / 21
Misclassification error rate: 0.125 = 3 / 24
> summary (prunedLenseTree4)

Classification tree:
snip.tree(tree = lenseTree, nodes = 6L)
Number of terminal nodes:  5
Residual mean deviance:  0.2846 = 5.407 / 19
Misclassification error rate: 0.04167 = 1 / 24
```

We can clearly see that the misclassification rate goes down as the number of nodes in the tree goes up, eventually reaching zero, when the tree has 7 nodes as we have seen earlier. As the number of nodes increases, the tree fits better with the training data, eventually fitting with it 100% by making as many nodes as possible. The corresponding trees are given in Figure 3.13. What these examples show is that bigger trees fit better to the training data as we know already. However, since we are using the entire dataset to build the trees every time, we do not know yet how to pick the best tree, since no validation or testing of the trees has been performed. Such validation is necessary on data previously unseen to prune a tree.

3.5.1.1   Cross-validation for Best Pruned Tree

Instead of trying out various values of tree sizes, and picking the best based on criteria that are yet to be decided, we can use a function provided by the

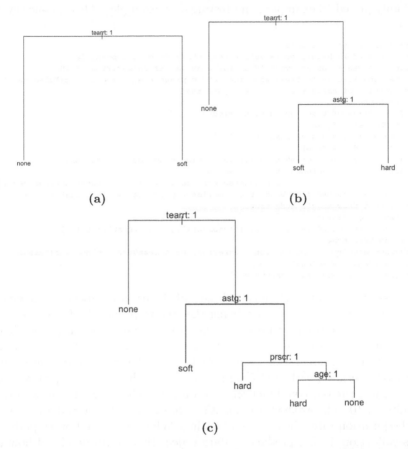

FIGURE 3.13: Pruned trees for the Contact Lense dataset. These trees have been obtained by pruning the tree in Figure 3.8 (a) Pruned down to 2 nodes. (b) Pruned down to 3 nodes. (c) Pruned down to 5 nodes.

`tree` library to pick the best size. It is the `cv.tree` function. The approach requires us to perform cross-validation, an approach to build machine learning models that we have seen in Chapter 2 in the context of regression. It divides the training set to a number of equal random disjoint parts, called folds. For example, it can break up the training set into, say three random folds or subsets, train on two of the subsets and test on the third. This is repeated a few times to find the "best" tree. For each tree it builds, it computes a metric that keeps track of misclassification rate, and picks the tree with the smallest rate.

The following piece of code using the `tree` library in R performs cross-validation, using a fold size of 3. Note that for the Lenses dataset with only 24 examples, each fold is of size 8. Thus, the training of the tree happens on randomly picked 16 examples, and testing on 8 examples. This is done several times.

```
1   Prune using cross-validation
2   #Here, we perform K-fold cross-validation, other folds are possible
3   #K can be given as an argument 'K=3' for example; if unspecified, K=10
4   cvPrunedLenseTree <- cv.tree(lenseTree, FUN = prune.tree,  method = c("misclass"))
5   #Find details of cross-validation was performed
6   cvPrunedLenseTree
7   #Find the sizes of trees which were tried
8   cvPrunedLenseTree$size
9   #Plot details of cross-validation analysis
10  plot (cvPrunedLenseTree)
11  #Find the tree sizes for which cross-validation produced best answers
12  bestSizes <- cvPrunedLenseTree$size [which
13                          (cvPrunedLenseTree$dev == min (cvPrunedLenseTree$dev))]
14  #The sizes are sorted from biggest to smallest, pick the smallest that is > 1
15  #plot cannot draw single node trees
16  l <- length(bestSizes)
17  bestSize <- if (bestSizes [1] > 1) {bestSizes[1]} else {bestSizes[1-1]}
18  #Draw the best tree
19  bestPrunedLenseTree <- prune.tree (lenseTree, best=bestSize, method = c("misclass"))
20  plot (bestPrunedLenseTree)
21  text (bestPrunedLenseTree, pretty = 0)
```

Cross-validation is performed in Line 4 of the program above. `cv.tree` can take several arguments, the first being the tree to prune, the FUN argument is a function that is used to perform the pruning, and the K argument being the number of folds to use for cross-validation. In this case, we prune the full tree, called here `LensesTree`, built in Section 3.4. The `tree` library provides a couple of functions for the FUN argument, one of which is called `prune.tree`, but we can write our own function if we want to. The value of K, if not given, defaults to 10. The `method` argument's value here tells the program to use misclassification rate when cross-validating. When the `tree` library performs cross-validation, it also produces a data object that stores details of how the pruning was performed. The summary shows the sizes of the pruned trees in terms of number of leaf nodes, the misclassification error (here, still called `dev` for deviance, which is the default method for error calculation) and the value of a cost-complexity metric k that we have not discussed yet.

```
> cvPrunedLenseTree
$size
[1] 7 5 3 1

$dev
[1] 10 10 10 12

$k
[1] -Inf  0.5  1.0  3.0

$method
[1] "misclass"

attr(,"class")
```

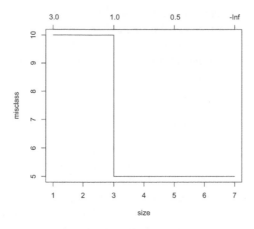

FIGURE 3.14: Plotting results of Cross-Validation Analysis for Trees using the `cv.tree` Function in the `tree` library.

The `size` vector's value says that the values of 7, 5, 3 and 1 were tried for the number of leaf nodes for the pruned tree. The `dev` values, in this case, tell us the misclassification rate, instead of deviance, which is the default method for error calculation, if `"misclass"` is mentioned as the `method` argument's value to `cv.tree`. In this case, we see that tree sizes of 7, 5 and 3 give a misclassification rate of 10% each and so any of these sizes can be used to draw the best cross-validated tree. We decide to pick the smallest of the sizes, if there are several, and draw a tree with that many leaf nodes. Since the `plot` function gives an error if we want to draw a tree of size 1, i.e., a stump only tree, we pick the smallest value that is bigger than one from the `cvPrunedtree$size` vector. The manner in which the misclassification rate varies as the tree size changes can be plotted using the

```
plot (cvPrunedLenseTree)
```

command. This plot is show in Figure 3.14 To obtain the best tree, we prune the original tree to the size of the best tree. This happens to be the 3-node tree in Figure 3.13 (b). The cross-validation analysis shows that the misclassification error comes down from 10 to 5 when the number of leaves becomes 3. Building bigger trees does not lower the error further.

3.5.2 Converting Trees to Rules

A decision tree can be easily converted to a set of if-then rules. In fact, each path from the root to a leaf node of a decision tree can be described by writing a corresponding if-then rule, with all the conditions traversed ANDed together. All the rules corresponding to all the paths can be written as a set of rules or a *ruleset*. Rulesets are generally easier to understand than trees since each rule describes a specific context associated with a class. Furthermore, a ruleset generated from a tree can be pruned to have fewer rules than than the tree has leaves, aiding comprehensibility. Converting to a set of rules removes the distinction between nodes that occur at upper levels of the tree and those that occur below because all of a rule's conditions are considered to be at the same level.

3.5.2.1   Building Rules in R

The `tree` library in R we have been discussing uses the Gini index to build decision trees. There are several libraries in R that can build trees. Another such library is the `C50` library that builds tree using the C5.0 algorithm [26]. This algorithm, which is a modified version of the C4.5 algorithm [25], builds decision trees using the entropy split criterion discussed earlier. The entropy and Gini index are not very different from each other in terms of behavior as we have seen in this chapter so far. We do not want to discuss the `C50` library here much, except to show that this library has the functionality to convert trees built into rules.

Consider the program below, which is like the programs we have written with the `tree` library, but with the `C50` library.

```
1   #Building decision tree with C5.0 by Quinlan (1993)
2   library (C50)
3
4   contactLenses <- read_csv("Datasets/txt/contactLenses.txt")
5   View(contactLenses)
6   #R reads the column values as strings
7   #We need to convert the strings into factors, i.e., enumerated values
8   contactLenses$age <- as.factor (contactLenses$age)
9   contactLenses$prscr <- as.factor (contactLenses$prscr)
10  contactLenses$astg <- as.factor(contactLenses$astg)
11  contactLenses$tearrt <- as.factor (contactLenses$tearrt)
12  contactLenses$class <- as.factor (contactLenses$class)
13
14  lenseTreeC50 <- C5.0(formula = contactLenses$class ~ . , data = contactLenses)
15  #plots a nice looking tree; not sure what plot function it used
16  plot(lenseTreeC50)
17  #Draw a tree that is pruned by default?
18  lenseTreeC50 <- C5.0(contactLenses$class ~ . , data = contactLenses)
19  plot(lenseTreeC50)
20
21  #obtain the rules
22  lenseRulesC50 <- C5.0(contactLenses$class ~ . , data = contactLenses,
23                                           rules = TRUE)
24  summary (lenseRulesC50)
```

The tree produced by the C5.0 library when plotted is shown in Figure 3.15. It looks a bit different since it is plotted by a different library. In this plot, there are several tree plotting functions and libraries, each doing it slightly differently. The leaves show the distribution of examples of various classes that they correspond to. The tree has been optimized in this case by pruning. A leaf node specifies how many examples are associated with it, in addition to the distribution of the various classes in these examples. For example, `Node` 2 has 12 examples associated with it, and all of the examples belong to the class **none**. `Node` 4 has 6 examples associated with it, with 4 in the **soft** class

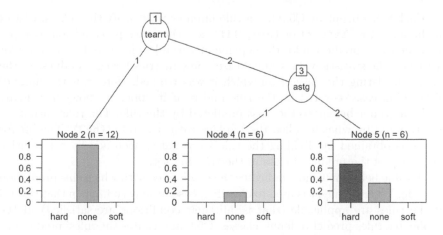

FIGURE 3.15: The Iris tree plotted by the C5.0 library in R.

and 2 in the **none** class with a label **soft** on it, with some error in it. For a leaf node in the decision tree, C5.0 outputs a confidence level computed as

$$\frac{n_{correct}}{n_{total}} \tag{3.10}$$

where $n_{correct}$ is the total number of training cases correctly classified by the leaf node, and n_{total} is the total number of training cases covered by the leaf node.

The command on line 22 obtains the rules. The last command prints the rules among other things. In this case, it prints the following rules.

```
Rule 1: (6/2, lift 3.8)
        astg = 2
        tearrt = 2
        -> class hard   [0.625]

Rule 2: (12, lift 1.5)
        tearrt = 1
        -> class none   [0.929]

Rule 3: (6/1, lift 3.6)
        astg = 1
        tearrt = 2
        -> class soft   [0.750]

Default class: none
```

The first rule can be written more readably as

```
If (astigmatism == yes) and (tearProductionRate  == normal)
        then class == hard
```

Each rule output by C5.0 has a rule number to identify the rule; statistics of the form (n, lift x) or (n/m, lift x); the rule's preconditions one per line; the rule's prediction for the dependent variable; and a confidence level for the rule. The statistic values summarize a few metrics for the goodness of the rule, considering the dataset on which it was trained. Here, n is the number of training cases covered by the rule and m, if it appears, shows how many of them do not belong to the class predicted by the rule. The rule's accuracy is estimated by what is called the Laplace ratio (n-m+1)/(n+2). The lift for a rule is obtained by dividing the rule's estimated accuracy by the relative frequency of the predicted class in the training set.

A value between 0 and 1 indicates the confidence with which this prediction is made. When a ruleset is used to classify a case, it may so happen that several of the rules are applicable (that is, all their conditions are satisfied). If the applicable rules predict different classes, C5.0 aggregates the rules' predictions to reach a verdict. In other words, each applicable rule votes for its predicted class with a voting weight equal to its confidence value, the votes are added, and the class with the highest total vote is chosen as the final prediction. There is also a default class, here *negative*, that is used when none of the rules apply.

The confidence level output by C5.0 is computed in two different ways for two different situations. For a ruleset, the confidence value is computed as

$$\frac{n_{correct} + 1}{n_{total} + 2} \tag{3.11}$$

where $n_{correct}$ is the number of training cases correctly classified by the rule, and n_{total} is the total training cases covered by the rule.

3.6 Evaluation of Classification

To build a decision tree or any classification model for that matter, we need to follow a certain protocol for training and testing so that experiments are performed consistently and report acceptable results.

3.6.1 Training and Testing Protocols

In some cases, such as machine learning contests that are sponsored by various conferences or organizations, we are given a training dataset only. We have to develop classification models and submit them to the contest organizers who run them on test datasets that are known only to them. This is to ensure that the contests are fair and no participant has unfair advantage. In these situations, the given training set must be divided into subsets following some protocol so that we can produce the best classifier for the task at hand.

In other cases, we are given an undifferentiated dataset to begin with. In such a case, we can randomly sample a sufficiently high percentage (say, 75% or 80%) of the data and call it the *training* (sub)set. The rest is the *testing* (sub)set. We train our model on this training set. Once we have built a tree or the classification model based on the training set, we run it on the test set and report results. This involves one time training and one time testing.

However, we know from our discussion so far in this chapter that when we build decision trees several times by training on sampled subsets of data, different trees are likely to be produced by the different samples using the same tree building algorithm. So, an approach to building a good decision tree should sample the data set a few times to build a number of decision trees, compute quality metrics on them, and possibly pick the one that gives the best evaluation. Thus, we can perform sampling of the dataset to obtain the training set a number of times. This leads to a protocol called *cross-validation*. We have seen a similar approach in building regression models in the previous chapter. We perform training and testing k times, each time testing on a (sub)set after training on the remaining subsets. The model obtained that performs the best is the one that is selected for reporting. To pick, say 75% of a dataset for training, we can think of picking four equal random sub-samples of data, each one being 25% of the dataset. We can then pick three of the subsamples as the training set, and the last one the test set. This process of picking four subsets can be done four times. Thus, the usual approach is to randomly divide up the dataset into k equal parts. The machine learning algorithm is trained on randomly selected $k - 1$ parts and tested on the remaining part. The results are recorded. We then perform the experiment again by picking another one of the k parts to test after training. We do not repeat the same part for testing. The purpose for doing cross-validation is to remove accidental regularities that may occur in a dataset by coincidence and gain some feel for how a machine learning algorithm is going to behave with unseen examples or in other words, gauge its generalized predictive capability. Since we pick the examples and folds randomly, the assumption is that any random or accidental associations or correlations among data items is at least partially removed from consideration.

Another commonly used protocol involves using what is called a validation (sub)set also. In other words, we sample the original dataset into three parts, say at 60%, 20% and 20%. The first part is the training (sub)set, the second is called the *validation (sub)set* and the third is the *testing (sub)set* we have seen earlier. In more general terms, we divide the dataset into k folds or parts, pick $k - 2$ parts for training, and one part for validating. We set aside a part for testing later. The training set is used to build the classification model, e.g., a decision tree. A classification model has parameters. For example, a tree may use the splitting approach, using Gini or Entropy as a parameter. The minimum number of data points at a node may be another parameter. Such parameters are changed and various results compared using the validation set. All combinations of training and validating are exhaustively experimented with to obtain the parameter values to give the best results. Once we have a validated classification model that we like, i.e., it has the best results so far, we use it as the final model. We run this final model on the test set and report results. Variations of the protocol are possible. For example, one can repeat the train-validate loop $k - 1$ times for every possible choice of the test set.

3.6.2　Evaluation Metrics

When we build a tree or a classification model from a training set, we need to evaluate it using one or more metrics. There are many metrics that have been used to evaluate decision trees in particular, and classification models in general. We have encountered a few of them already in this chapter. We discuss several relevant metrics in one single place in this section.

The commonly used metrics for evaluating a classification algorithm are *Accuracy*, *Precision*, *Recall* and *F-measure*. We will provide the definitions below. We have already seen the use of metrics such as deviance and misclassification rate earlier in the chapter.

We use the following notations for our discussion of these metrics. These notations are used in defining the metrics later. To start, assume that we are performing binary classification, i.e., we have two classes. These could be *TRUE* and *FALSE*, *yes* and *no*, *PlayTenns* and *notPlayTennis*, etc.

3.6.2.1　Basic Terminologies

True Positives: True positives are positive examples of a class in the test set that are classified as positive by the trained program. The positive class is the *TRUE*, *yes* or *PlayTennis* class in the examples above. The term *TP* is the count the number of true positive examples.

False Negatives: This is the set of positive examples in the test set that are classified as negative by the trained program. The negative class is the *FALSE*, *no* or *notPlayTennis* class in the examples above. The abbreviation *FN* is used to record the count of true negative examples. For example, it is the number of *PlayTennis* cases that are classified as *notPlayTennis* by the trained classifier, in the case of the classifier discussed in this section.

True Negatives: This is the set of negative examples in the test set that are classified as negative by the trained program. For example, it is the set of *notPlayTennis* examples that are correctly classified as *notPlayTennis*. *TN* is the size of this set.

False Positives: This is the set of negative examples in the test set that are classified as positive by the trained program. For example, *FP* is the number of *notPlayTennis* examples that are classified as *PlayTennis*.

In binary classification, these four numbers can be displayed in a 2×2 matrix form. This matrix is called the *Confusion Matrix*. We see such a matrix in Table 3.7.

TABLE 3.7: Binary Confusion Table

Confusion Matrix		True Condition	
		Positive	**Negative**
Predicted Condition	Positive	TP	FN
	Negative	FP	TN

3.6.2.2 Metrics for Binary Classification

We first discuss the most commonly used terms for evaluating binary classification models in machine learning. There is a plethora of terms, quite frequently with different terms used to describe the same metric by different groups of practitioners and researchers. A metric that seems to be used most commonly in machine learning literature is Accuracy. Accuracy is computed by most machine learning libraries.

Accuracy is defined as the percentage of unseen examples classified correctly by a classifier. In other words,

$$Accuracy = \frac{TP + TN}{TP + TN + FN + FP}, \tag{3.12}$$

using the basic terminologies, defined above. The numerator contains the total number of test examples that were classified correctly by the trained classifier. The denominator contains the total number of test examples. $TP + TN$ is the total number of examples classified as positive, and $FN + FP$ is the total number of examples that are classified as negative. $TP + FN$ is sometimes called P, the count of positive examples in the training set. $TN + FP$ is called N sometimes, the total count of negative examples in the training set. We note that if accuracy is 100%, all the examples are classified correctly, and everything is perfect. If the classes are balanced, i.e., the number of examples from the two classes are approximately equal in the test set, accuracy works well as a metric for the goodness of a classifier. However, if the classes are unbalanced, i.e., one class has more examples in the test set than the other class, a high accuracy may not indicate that the results are actually good. For example, if 90% of our examples in the test set are positive and only 10% are negative, we can achieve 90% accuracy by simply classifying each test example as positive, without doing any work. This happens even though we are 100% wrong in classifying the negatives.

Thus, in addition to or in lieu of accuracy, it is common to calculate a few other metrics for evaluation of classifiers. A few common ones are precision, recall and F-measure. For our discussion below, let us call the two classes simply + and −. We can compute precision and recall for the + class using the definitions given below.

$$Precision_+ = \frac{TP}{TP + FP} \qquad (3.13)$$

$$Recall_+ = \frac{TP}{TP + FN} \qquad (3.14)$$

$Precision_+$ gives a measure of precision for the + class. It is traditionally called just *precision*. In statistics, it is also called the *Positive Predictive Value* or *PPV*. Precision is measured by dividing the number of test examples that are correctly classified as + by the classifier by the number of test examples that are classified as + (either correctly or wrongly) by the classifier. Thus, precision for class + is the proportion of all test examples that have been classified as belonging to class + that are true members of class +. For example, suppose our trained classifier classifies 10 (= $TP + FP = P$) test examples as belonging to the + class. Suppose 8 (= TP) of these are correctly classified and 2 (= FP) are wrongly classified. In this situation, the precision for the yes class is $\frac{8}{10}$ or 80%.

Recall measures the proportion of test examples belonging to a certain class that are classified correctly as belonging to the class. The formula given above for recall computes recall for the + class. This is simply called *recall* for binary classification. It is also called *True Positive Rate* or *TPR* in statistics. Assume we have a total of ten examples of the + class in the test set. Of these, our trained classifier classifies eight as belonging to the + class. Then, our recall for the + class, or simply recall is $\frac{8}{10} = .8$ of 80%.

If we use the terms *Precision* and *Recall* without a subscript, they stand for the values of precision and recall for the + class.

When there are two measures for a class, some find it difficult to compare results of learning experiments. So, a combined metric for a class is used by many researchers. The measure normally used is called the F-measure. It is defined as follows.

$$\frac{1}{F} = \frac{1}{2}\left(\frac{1}{Precision} + \frac{1}{Recall}\right) \qquad (3.15)$$

or

$$F = 2\frac{Precision \times Recall}{Precision + Recall} \qquad (3.16)$$

Here *Precision* stands for $Precision_+$ and *Recall* stands for $Recall_+$. Thus, the F-measure for a class is the harmonic mean of the recall and precision for the class. When the F-measure is reported, we have one value to evaluate how our classifier learned to classify, considering the + class only. The most commonly used metrics for classification in machine learning are given in Table 3.8.

3.6.2.3 Evaluation of Binary Classification Using R

All classification libraries report some evaluation metrics as we have seen with the tree libraries we have discussed here. There is a useful library called

TABLE 3.8: Common evaluation metrics for classification models in Machine Learning

Our terminology	Commonly used terminology	Commonly used terminology in statistics	Formula
Precision$_+$	Precision	Positive Predictive Value (PPV)	$\frac{TP}{TP+FP}$
Recall$_+$	Recall	True Positive Rate (TPR), Sensitivity	$\frac{TP}{TP+FN}$
F	F_1-measure		$2\frac{Precision \times Recall}{Precision+Recall}$
Accuracy	Accuracy	Accuracy	$\frac{TP+TN}{TP+TN+FN+FP}$

caret that produces detailed evaluation reports. There is another library called yardstick that also produces detailed reports. Below, we show how caret returns evaluation metrics.

```
1  library (caret) #for confusionMatrix
2  library (tree)  #to build trees
3  library (readr) #for read_csv
4
5  weather <- read_csv("Datasets/txt/weather.numeric.txt")
6  View(weather)
7
8  #R reads the column values as strings
9  #We need to convert the class column into factor, i.e., enumerated values
10 weather$outlook <- as.factor (weather$outlook)
11 weather$windy <- as.factor (weather$windy)
12 weather$play <- as.factor (weather$play)
13
14 #set a seed to generate random numbers
15 set.seed(2)
16 #70% training and 30% testing, because our dataset is very small
17 trainIndex <- caret::createDataPartition (y=weather$play, p=0.7, list = FALSE)
18 weatherTrainingSet <- weather [trainIndex,] #obtain the training set
19 weatherTestSet <- weather [-trainIndex,]    #obtain the test set
20
21 #Build all the way to the bottom till every node is homogeneous
22 #minsize is the minimum number of data points at a node
23 weatherTree <- tree (formula = weather$play ~ ., data = weather,
24                                   split = c("gini"), minsize = 1)
25 weatherTree <- prune.tress (weatherTree, method = c ("misclass"), best = 3)
26 summary (weatherTree)
27
28 #Plot the tree
29 plot (weatherTree)
30 text (weatherTree, pretty = 0)
31
32 #obtain predictions for the test set
33 #Gives probabilities for the two classes: yes, no
34 weatherPredictions <- predict (weatherTree, newdata = weatherTestSet)
35 #actual labels for the classes in weatherTestSet
36 weatherTestLabels <- weatherTestSet$play
37
38 #obtain weatherPredLabels as the predicted labels
39 weatherPredLabels <- vector(mode="character", length = NROW(weatherTestLabels))
40 for (i in 1:NROW (weatherPredictions)){
41     if (weatherPredictions[i,1] == 1){
```

```
42            weatherPredLabels [i] <- "yes"
43        }
44       if (weatherPredictions[i,2] == 1){
45            weatherPredLabels [i] <- "no"
46        }
47 }
48 weatherPredLabels <- as.factor (weatherPredLabels)
49
50 #prints out confusion matrix and other details
51 caret::confusionMatrix(weatherPredLabels, weatherTestLabels)
```

We use the `caret` library in this program. Using this library, we divide up a dataset into training and testing (sub)sets, say with a 70% and 30% split in lines 17-19. Once we have done the splitting, we train the program on the training set. We test it by asking the trained model to predict the classification of the testing set. This results are in `weatherPredictions`, which is a matrix like the following.

```
   no yes
1   0   1
2   1   0
3   0   1
```

We process this matrix to obtain a vector that has the predicted labels for the test set. In this case, the test set has only three examples. We obtain what is called `weatherPredLabels` which contains the following.

```
[1] no   yes no
```

So, when we use the `caret::confusionMatrix()` to obtain results of the classification, we obtain the following.

```
              Reference
Confusion Matrix and Statistics

              Reference
Prediction no yes
       no    1   1
       yes   0   1

                     Accuracy : 0.6667
                       95% CI : (0.0943, 0.9916)
          No Information Rate : 0.6667
          P-Value [Acc > NIR] : 0.7407

                        Kappa : 0.4
       Mcnemar's Test P-Value : 1.0000

                  Sensitivity : 1.0000
                  Specificity : 0.5000
               Pos Pred Value : 0.5000
               Neg Pred Value : 1.0000
```

```
          Prevalence : 0.3333
      Detection Rate : 0.3333
Detection Prevalence : 0.6667
   Balanced Accuracy : 0.7500

   'Positive' Class : no
```

The accuracy value for this classifier is 0.6667. As see in Table 3.8, Sensitivity is the same as Recall, and Positive Predictive Value is the same as Precision. R uses names preferred by statisticians since it is primarily a programming language for statistics. It does not provide a value for F-measure, but it can be easily calculated if Precision and Recall are known. In this case, with only three test examples, the metric values we obtain are quite bad. But, it gives us a good idea the kinds of evaluations we can perform. Note that with training on 70% of the data, the tree obtained is the same as in Figure 3.16.

Additional Metrics for Binary Classification: In addition to accuracy, precision, recall and F-measure that are commonly used in machine learning, several other terms are also used, especially in statistics and also in some sub-areas of machine learning. Many software libraries in R and other languages provide some of these metrics also. We see that the `caret` library returns a number of additional metrics also. We do not discuss these metrics in detail, but provide Table 3.9 with brief descriptions: *Negative Predictive Value, Positive Predictive Value, Kappa, Prevalence, Detection Rate, Detection Prevalence, Balanced Accuracy.*

3.6.2.4 Metrics for Multi-class Classification

If we have more than two classes, say k classes in total, we have to compute metrics for each of the classes. When individual results are reported for a number of classes, we can obtain good insights regarding how the classifier performs as far as various classes go. This may spur us to find classifiers that work better for the underperforming classes. However, if we obtain a number of individual metric values for a number of classes, it may be difficult to get an overall feel for the performance of the classifier. If we have k classes, even if we focus on F-measure only, we will have k different F-measure values. In

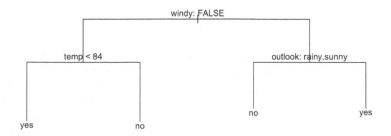

FIGURE 3.16: PlayTennis Tree with Confusion Matrix.

TABLE 3.9:   Additional evaluation metrics for Classification models in Machine Learning

Terminology	Formula	Comments
Negative Predictive Value (NPV)	$\frac{TN}{TN+FN}$	It can be thought of as *Precision*_ or precision for the -ve class, i.e., the proportion of examples returned as -ve are actually -ve
True Negative Rate (TNR), Specificity	$\frac{TN}{TN+FP}$	It can be thought of as *Recall*_, i.e., the proportion of all -ve examples known that are classified as -ve
Balanced Accuracy	$\frac{1}{2}\left(\frac{TP}{TP+FN} + \frac{TN}{FP+TN}\right)$	It is equal to $\frac{1}{2}(Sensitivity + Specificity)$. It is a better metric compared to accuracy if the classes are not balanced.
Prevalence	$\frac{TP+FN}{TP+FP+TN+FN}$	Percentage of test examples that are +ve, i.e., how prevalent or common is the +ve class in the test dataset
Detection Rate	$\frac{TP}{TP+FP+TN+FN}$	It is the proportion of examples classified as +ve to the size of the test set.
Detection Prevalence	$\frac{TP+FN}{TP+FP+TN+FN}$	It is the proportion of examples that are classified as +ve to the size of the test dataset.
Kappa	$\frac{Accuracy_{Observed} - Accuracy_{Expected}}{1 - Accuracy_{Expected}}$ where $Accuracy_{Expected}$ is the average of 1 (the accuracy of the labels given by expert), and the accuracy of the machine learned classifier $Accuracy_{Observed}$	Kappa or Cohen's Kappa is like classification accuracy, but can be thought of as a comparison of overall accuracy to expected accuracy of a random classifier. A kappa value of 0.75 or above is usually considered good for a classifier.

(*Continued on next page*)

TABLE 3.9: (*Continued*)

Terminology	Formula	Comments
McNemar's Test P-value	Compute $\frac{(FN-FP)^2}{FN+FP}$ with the assumption that it follows the χ^2 distribution with 1 degree freedom. The null hypothesis is that probability (Predicted Positives) = probability (True Positives).	A statistical computation that says if the classifier for the +ve class is as good as the classifier for the -ve class.
No Information Rate (NIR)	If y_i and $\hat{y}_i$ are the true labels and predicted labels for N test examples, respectively, $NIR = \sum_{i=1}^{N} \sum_{j=1}^{N} (y_i \neq \hat{y}_j)$	The proportion of classes that are correct when labels are randomly guessed on all test examples. A value of less than 0.05 tells us if the classifier is better than NIR classifier

such situations, one can obtain an average of the F-measure values as a single metric value for the overall quality of the classifier.

We can obtain the average in two ways: macro-averaging and micro-averaging.

Macro-averaging over All Classes: In macro-averaging, the metric values (e.g., values of precision, recall and F-measure) are calculated separately for each class. A macro-averaged metric is simply the average of the metric values for each class. If we have k classes, we compute the sum of the precision values for all classes and divide the sum by k to obtain macro-averaged precision. We compute macro-averaged precision, recall and F-measure in this fashion. In macro-averaging, we treat each class as equal, even if one class has many more examples than another.

Micro-averaging over All Test Examples: In micro-averaging, we obtain the precision, recall and F-measure values for each individual class as usual. However, when we average to obtain the overall result, we perform a weighted average computation. In other words, we weigh the value for a class by the proportion of total number of examples the class contains. For example, if we have two classes, yes (+) and no (−), the micro-averaged precision can be obtained as

$$Precision_{micro} = p_+ \times precision_+ + p_- \times precision_- \qquad (3.17)$$

where p_+ and p_- are the proportions of elements in the test set that are in the + and − classes, respectively. Obviously, $p_+ = 1 - p_-$. Here, $precision_{micro}$ gives us an idea how the trained classifier performs in regards to precision over the two classes.

If we have k classes, we compute the micro-averaged metric by performing a weighted sum of the metric values for the k classes. Usually, if the various classes are not balanced, the proportionally larger sized classes contribute more to the final micro-averaged measure. Thus, it is preferred if this is the case. If the classes are more or less balanced, either one of micro and macro averaging works.

Multi-class Classification in R: We can easily change the R program we presented in the previous section for multi-class classification to work with the Iris dataset, which has three classes: setosa, versicolor and virginica. We do not show the program here for lack of space, but we show the results that are produced by a modified program like this.

```
 1 │ Confusion Matrix and Statistics
 2 │
 3 │               Reference
 4 │ Prediction    setosa versicolor virginica
 5 │   setosa         10         2         9
 6 │   versicolor      0         8         1
 7 │   virginica       0         0         0
 8 │
 9 │ Overall Statistics
10 │
11 │                 Accuracy : 0.6
12 │                   95% CI : (0.406, 0.7734)
13 │      No Information Rate : 0.3333
14 │      P-Value [Acc > NIR] : 0.002458
15 │
16 │                    Kappa : 0.4
17 │   Mcnemar's Test P-Value : 0.007383
18 │
19 │ Statistics by Class:
20 │
21 │                        Class: setosa Class: versicolor Class: virginica
22 │ Sensitivity                   1.0000            0.8000           0.0000
23 │ Specificity                   0.4500            0.9500           1.0000
24 │ Pos Pred Value                0.4762            0.8889              NaN
25 │ Neg Pred Value                1.0000            0.9048           0.6667
26 │ Prevalence                    0.3333            0.3333           0.3333
27 │ Detection Rate                0.3333            0.2667           0.0000
28 │ Detection Prevalence          0.7000            0.3000           0.0000
29 │ Balanced Accuracy             0.7250            0.8750           0.5000
```

We see that each of the metrics has been computed separately for each of the classes by R's `caret` library. We already know that the term used by R for precision is Positive Predictive Value, and for recall, the corresponding term used by `caret` is Sensitivity. Thus, in the results returned by R above, we see the values of these two metrics for each of the three classes: `setosa`, `versicolor` and `virginica`. However, unfortunately, the `caret` library does not compute the value of F-measure. We can write a program that can compute the F-measure for each of the classes separately, and then obtain the macro- and micro-averaged F-measure for the entire classifier.

We see in the results that precision for the virginica class is undefined or Nan, i.e., not a number. This can happen only if $TP + FP = 0$; in other words, nothing was classified as virginica, either rightly or wrongly. If we look at the confusion matrix, that is what we see. In such a case, the program cannot compute F-measure since the denominator has the sum precision and recall, and this sum happens to be zero in this case.

There are R libraries that compute macro- and micro- F-measures and other metrics. One such library is `yardstick`. We do not discuss this library here since the `caret` library is more general and has a lot more functionality than what we have touched upon here. However, if it is necessary to compute these metrics, one should use `yardstick`, possibly in addition to `caret`.

3.7 Ensemble Classification and Random Forests

In general parlance, an ensemble is a collection of musicians, actors or dancers that perform not as individuals, but as a group. Usually, in an ensemble, there is differentiation among the participants, although it is common to have sub-groups of participants that are similar. For example, in an orchestra, which is an ensemble, it is likely that we have sub-groups of string instruments, percussion instruments, brass instruments and woodwinds. We may have further sub-groups within these sub-groups. The idea in a classical music orchestra ensemble is to have a variety of participants that are essentially all creating music, but in different ways, and when everything works together, it does so beautifully, creating entertaining and memorable music. The same is true of an ensemble of actors or dancers as well, in their own domains.

The concept of an ensemble is used in machine learning also, particularly in classification. We have seen decision tree classifiers in this chapter. There are other types of classifiers including support vector machines, naive Bayes classifiers and artificial neural networks. Each classifier has its own inductive bias, and its strengths and weaknesses. We can create an ensemble of classifiers if we have several classifiers, each possibly of a different type, train these classifiers separately, and then use the trained classifiers to classify unseen examples. The final classification is obtained by combining the classifications of the individual classifiers, using a policy like majority voting, weighted voting, or even consensus.

3.7.1 Bootstrapping and Bagging (Bootstrapped Aggregation)

It is also possible to build an ensemble from a number of participants that are similar to each other, but still sufficiently different. This may not be an ensemble in the traditional view, but if the participants are at least a bit different from each other, the ensemble could still be valuable or useful. This is what happens when we make an ensemble of decision trees. For example, given a training dataset with N examples, we can create a number of derived datasets by randomly sampling with replacement, the original training dataset many times. This is called bootstrap aggregation or *bagging*. Bagging attempts to remove accidental correlations that may be present in the training dataset, and a set of bagged decision trees usually performs better than a single decision tree. In fact, a single decision tree is usually not a very accurate classifier for predicting classes of unknown examples. In addition, a single decision tree is often unstable with high variance, i.e., slight change in the training examples may change the actual tree quite a bit, even if it is pruned or cross-validated. That is why a set of bagged decision trees often has better predictive ability than a single decision tree. Below, we discuss the idea of bootstrapping clearly.

Let us denote the training set as

$$\mathbf{Z} = \left\{ z^{(i)}, \cdots, \vec{z}^{(N)} \right\}.$$

We have N data examples. $\vec{x}^{(i)}$ is the ith example and $y^{(i)}$ is its label. Thus,

$$\vec{z}^{(i)} = \left\langle \vec{x}^{(i)}, y^{(i)} \right\rangle.$$

In bootstrapping, we randomly create B (say, 100) datasets from the original dataset $\mathbf{Z}$ by picking elements randomly with replacement. Each dataset $\mathbf{Z}^{(b)}, b = 1, \cdots, B$ contains N examples like the original dataset $\mathbf{Z}$, but an example can occur a number of times in a derived or bootstrap dataset. We create a set of bootstrap datasets:

$$\mathcal{Z} = \left\{ \mathbf{Z}^{(1)}, \cdots, \mathbf{Z}^{(B)} \right\}.$$

The metric in the figure can be any metric such as accuracy, precision, recall or F-measure.

Each of the B classifiers is, in fact, a hypothesis regarding how the ideal classifier should work. A single decision tree is, in some sense, a weak classifier since it has high variance in producing results, and thus, the results vary from tree to tree considerably. Let the bootstrap classifier $\mathcal{H}(\vec{x})$ be written as

$$\mathcal{H}(\vec{x}) = \left\{ h^{(1)}(\vec{x}), \cdots, h^{(K)}(\vec{x}) \right\}. \tag{3.18}$$

Here, each $h^{(k)}(\vec{x})$ is a decision tree.

Let us assume our metric is accuracy. Since each bootstrap sample likely creates a different classifier, there will be variations in the classification results. A decision tree is a so-called high-variance classifier. In other words, there will be a lot of variations in the results, i.e., metric values, when different datasets are used to train a decision tree. Thus, bootstrap datasets are also likely to lead to high variations in the metric values. If we have B bootstrap datasets, we can compute the accuracy of the decision tree built from each dataset (using the same training and testing protocol) and then obtain the variance in accuracy using the standard formula.

$$\widehat{var}[acc(\mathbf{Z})] = \frac{1}{B-1} \sum_{b=1}^{B} \left(acc(\mathbf{Z}^{(b)}) - \overline{acc} \right)^2 \tag{3.19}$$

where

$$\overline{acc} = \frac{1}{B} \sum_{b=1}^{B} acc(\mathbf{Z}^{(b)})$$

and $\widehat{var}[acc(\mathbf{Z})]$ is the variance of $acc(\mathbf{Z})$ under sampling for the dataset $\mathbf{Z}$. This variance is likely to be high for decision trees. $acc(\mathbf{Z}^{(b)}$ is the accuracy obtained for the decision tree built from the bootstrap sample $\mathbf{Z}^{(b)}$, i.e., accuracy for classifier $h^{(b)}$.

Assuming that we have two classes in the dataset and we obtain bootstrap samples, the probability that a data example $\vec{z}^{(i)} = \langle x^{(i)}, y^{(i)} \rangle$ belongs to the bth bootstrap sample $\mathbf{Z}^{(b)}$ is

$$Prob(\vec{z}^{(i)} \in \mathbf{Z}^{(b)}) = \left(1 - \frac{1}{N}\right)^N \tag{3.20}$$

$$= e^{-1} = 0.368.$$

This is because a specific example $\vec{z}^{(i)}$ out of N can be picked in the first trial or picking with a probability of $\frac{1}{N}$. Therefore, the probability that a specific example is not picked in the first trial is $1 - \frac{1}{N}$. The probability that $\vec{z}^{(i)})$ is not picked in N pickings is

$$\left(1 - \frac{1}{N}\right)\left(1 - \frac{1}{N}\right) \cdots \left(1 - \frac{1}{N}\right)$$

with N multiplications or $\left(1 - \frac{1}{N}\right)^N$. Thus, the probability that a specific item $\vec{z}^{(i)})$ is not picked at all in a bootstrap sample of N pickings is $\left(1 - \frac{1}{N}\right)^N$. If $N \to \infty$, we know that $\left(1 - \frac{1}{N}\right)^N \to \frac{1}{e}$, where $e = 2.718$. In other words, a specific item is not picked with a probability of 0.368 in a bootstrap sample. Alternatively, a bootstrap sample contains only $1 - 0.368 = 0.632$ or 63.2% of the examples on average.

How do we compute the total error of a bagged set of trees? One way to do so is to look at all the trees produced in the bootstrap process and compute the error for every example. Thus, an estimate may be [11]:

$$\widehat{err}'_{boot} = \frac{1}{B}\frac{1}{N} \sum_{b=1}^{B} \sum_{i=1}^{N} err_{indiv}\left(y^{(i)}, h^{(b)}\left(\vec{x}^{(i)}\right)\right) \tag{3.21}$$

where $err_{indiv}(.)$ computes if an individual example $\vec{x}^{(i)}$ has been classified incorrectly, given that the actual label is $y^{(i)}$. $h^{(b)}()$ is the bth bootstrap decision tree. One possible $err_{indiv(,)}$ can simply be misclassification error $misclass(y^{(i)}, h^{(b)}(\vec{x}^{(i)}))$, which takes into account if an example $\vec{x}^{(i)}$ is classified properly. This is averaged over all B trees, each classifying N examples.

$$\widehat{err}'_{boot} = \frac{1}{B}\frac{1}{N} \sum_{b=1}^{B} \sum_{i=1}^{N} misclass\left(y^{(i)}, h^{(b)}\left(\vec{x}^{(i)}\right)\right) \tag{3.22}$$

Other individual error or loss functions can also be used.

In the formula for error above, it is not clear which $\vec{x}^{(i)}$s are considered in the computation of individual errors. In simplistic calculation, we can train on all the bootstrapped training sets, and test on the entire dataset. But, there is a considerable amount of overlap between each of the bootstrap samples and the original dataset, since each bootstrap sample contains about 63.2% of the original dataset. As we all know, training and testing on the same data examples is a terrible idea. The error computation can be done better, if we do the following. For each example $\vec{x}^{(i)}$ in the original dataset, find those bootstrap datasets indexed $-i$ in which the sample does not occur, and perform the individual error calculation only for those bootstrap datasets. Here, the -ve sign in front of the actual index i is to indicate the non-occurrence of $\vec{x}^{(i)}$ in the bootstrap dataset. The new formula for error becomes

$$\widehat{err}_{bootstrap} = \frac{1}{N} \sum_{i=1}^{N} \frac{1}{|C^{-i}|} \sum_{b \in C^{-i}} err_{indiv}\left(y^{(i)}, h^{(b)}\left(\vec{x}^{(i)}\right) \right). \tag{3.23}$$

What has changed from the previous formula is that instead of $\frac{1}{B}$ as a multiplier, we now have $\frac{1}{|C^{-i}|}$ as a multiplier, and that the summation inside has changed because we want to exclude those classifiers which have been trained on $\vec{x}^{(i)}$. This error is called the *Out Of Bag* (OOB) error for an ensembled set of classifiers. We can use misclassification error or any other kind of error for the individual error. This new error does not suffer from overfitting that occurs in the first formula. There are other improvements to the prediction error estimate that can be derived, but they become complex and we do not discuss them here. Interested readers can refer to [11].

Once a classifier has been trained, the classification of an example $\vec{x}^{(t)}$ is given as

$$\hat{f}^{(b)}\left(\vec{x}^{(t)}\right) = \arg\max_{c_i \in \mathcal{C}} \sum_{b=1}^{B} I\left(h^{(b)}\left(\vec{x}^{(t)}\right) == c_i \right) \tag{3.24}$$

where I is an indicator function whose value is 0 or 1. The indicator function takes a value of 1, if the condition given as argument is satisfied, 0 otherwise. The formula above essentially takes a vote of the B classifiers and the class c_i that gets the majority vote wins the classification race.

Bagging or Bootstrap Aggregation for classification can be understood in terms of joint committee decision coming from a number of independent "weak learners" [8]. This idea is motivated by by a heuristic that the collective judgement of a diverse and independent committee of people is usually better than the judgement of one single person. However, there is one main issue with bagged classifiers, which is that the bootstrap samples and hence the bagged trees are not independent of each other.

3.7.2 Feature Sampling and Random Forests

The idea of bagging has been modified to work even better, at least in the context of decision trees. This approach bags the training dataset to produce a number of randomly sampled datasets. In addition, when building a decision tree out of a single such sampled dataset $\mathbf{Z}^{(b)}$, it additionally samples the set of features $\vec{f}$ to form a sampled subset of features at every node. The reason for sampling the feature set is that the features may not be independent of each other; some features may be correlated to some other features. This sampling of features is likely to untangle some of the correlations that may exist among the features.

The number of features sampled at each node is usually quite small. It can be as small as 1 or 2, but for classification problems, a recommended value is $\sqrt{p}$ where p is the total number of features in an example. Thus, if an example is given in terms of 10 features, $\sqrt{10} \approx 3$, and it may be enough to pick 3 features to split the decision tree at a certain node [4]. Thus, for every tree that is built, all features F are at play, but at every node only a small number of features are picked to be used as candidates for splitting. If we follow Breiman, the random picking of features from all the features is done at every node. The process for splitting at a specific node remains the same as before (e.g., using entropy, Gini or variance), but only the picked features are considered as candidates. It is possible to take other approaches such as choosing a random subset of features first and use the subset to build an entire tree; for example, [16] did so.

Random forests are difficult to analyze mathematically. There have been attempts at analysis of random forests, but most have done so in the context of regression, where the dependent variable is numeric. We do not discuss any complex analysis here, except discuss an important finding in the next subsection.

In practice, the OOB error rate, discussed earlier in the context of bagging and a host of other metrics have been used as measures of goodness for random forests.

3.7.2.1 Do Random Forests Converge?

Assume we have a test sample $\vec{z} = \langle \vec{x}, y \rangle$. When B ensembled classifiers classify it, it is likely that some classify it correctly and others do not. Let us define the *individual margin* $\widehat{m}_{indiv}$ for an example as the proportion of classifiers that classify $\vec{x}$ correctly minus the proportion of classifiers that classify it as the next-best class [4].

$$\widehat{m}_{indiv}(\vec{z}) = \widehat{m}_{indiv}(\vec{x}, y) \tag{3.25}$$

$$= \frac{1}{B} \sum_{b=1}^{B} I\big(h^{(b)}(\vec{x}) == y\big) - max \, \frac{1}{B} \sum_{\substack{b=1 \\ y \prime \neq y}}^{B} \big(h^{(b)}(\vec{x}) == y\prime\big)$$

$$= \widehat{P}_b\big(h^{(b)}(\vec{x}) == y\big) - max \, \widehat{P}_{\substack{b=1 \\ y\prime \neq y}} \big(h^{(b)}(\vec{x}) == y\big)$$

In the second line above, the first expression counts the number of classifiers that classify $\vec{x}$ correctly, divided by B, the total number of classifiers. As explained earlier, $I()$ has a value of 1 if the condition given as argument holds, 0 if it does not hold. The numerator in the second expression in the second line starts by counting the number of classifiers that classify $\vec{x}$ wrongly; the *max* in front gives the number of classifiers that classify $\vec{x}$ as belonging to second-best class. The third line in the formula writes the proportion of classifiers that satisfy the condition as observed probability $\widehat{P}_b$, the subscript b saying it is over the B classifiers, the generic one indexed by b.

The formula above gives the individual margin for one test example $\vec{z} = \langle \vec{x}, y \rangle$. We can obtain the individual margins for all test examples, and get a metric for the entire set of classifiers over the entire dataset. Let us call this $\widehat{m}(\mathbf{Z})$ where $\mathbf{Z}$ is the entire test dataset, and $\widehat{m}$ is the empirical margin for the entire set of classifiers:

$$\widehat{m}(\mathbf{Z}) = \frac{1}{N} \sum_{i=1}^{N} m_{indiv}\left(\vec{z}^{(i)}\right). \tag{3.26}$$

This measures the extent to which the average number of votes for the correct class exceeds the average number of votes for the next-best class, considering all test examples and all ensembled classifiers.

According to Breiman, who invented random forests, the goal of creating an ensemble of classifiers was to optimize (maximize) the value of $\widehat{m}(\mathbf{Z})$. And, following our discussion earlier, we may consider changing it to reflect only those classifiers which are not trained on a test example $\vec{z} = \langle \vec{x}, y \rangle$, although Breiman's original paper does so. With Breiman's definition of $\widehat{m}(\mathbf{Z})$, we need to maximize its value for an ensemble. If we want to set it up as a minimization problem, we can start by writing up a condition: $\widehat{m}(\mathbf{Z}) < 0$, which says that the average empirical margin is less than 0, which is undesirable if our goal is to build a good ensemble. If the average empirical margin is less than 0, our ensemble is a bad one. We can write a loss or error function for the classifier ensemble as

$$\epsilon = \frac{1}{N}\widehat{m}(\mathbf{Z}) < 0 \tag{3.27}$$

i.e., the fraction of examples for which the empirical average margin is undesirable.

Considering this as an error function, Breiman was able to show that a random forest never overfits as more trees are added. In other words, he showed theoretically that as $B \to \infty$, the value of ϵ converges, providing a limiting error. He also was able to obtain a weak upper limit to the error. The theoretical proof is beyond the scope of this book, but suffice it to say that random forests do not overfit, and good experimental performance over many years have made random forest very popular classification algorithm.

3.7.3   Random Forests in R

There are several libraries in R that can compute random forests. We discuss the `randomForest` package here. It is simple to use and it implements the idea of random forests developed by [4, 3]. As default, it produces and ensemble of 500 trees. For classification, it takes the majority vote of the trees it generates to produce the final classification. Below, we use the library to produce a random forest for the `Iris` dataset.

```
1  #file: irisRandomForest.r
2  library (caret) #for confusionMatrix
3  library (randomForest)   #to build random forest
4  library (readr) #for read_csv
5
6  iris <- readr::read_csv("Datasets/txt/iris.txt")
7  View(iris)
8
9  #R reads the column values as strings; convert the class column into factor,
10 #i.e., enumerated values
11 iris$iType <- as.factor (iris$iType)
12
13 #set a seed to generate random numbers
14 set.seed(2)
15 #80% training and 20% testing
16 trainIndex <- caret::createDataPartition (y=iris$iType, p=0.8, list = FALSE)
17 irisTrainingSet <- iris [trainIndex,] #obtain the training set
18 irisTestSet <- iris [-trainIndex,]    #obtain the test set
19
20 #Build a random forest of 500 trees trained on the training set
21 irisForest <- randomForest (formula = irisTrainingSet$iType ~ .,
22                             data = irisTrainingSet)
23 plot (irisForest)
24 summary (irisForest)
25
26 #obtain predictions for the test set
27 #Gives probabilities for the three classes: setosa, versicolor, virginica
28 irisPredictions <- predict (irisForest, newdata = irisTestSet)
29 #actual labels for the classes in irisTestSet
30 irisTestLabels <- irisTestSet$iType
31
32 #prints out confusion matrix and other details
33 caret::confusionMatrix(irisPredictions, irisTestLabels)
```

As usual, in lines 1-17, we load the necessary libraries, load the dataset, create factors for the discrete variables, and create the partitions for training and testing. Line 20 produces the random forest, 500 trees strong by default. Line 22 plots the random forest. Plotting means plotting the error in classification against the number of trees. The plot is shown in Figure 3.17. The upper blue plot shows the testing error and the lower plot shows the training error. As normal, the training error is much lower than the testing error. Both errors fluctuate in the beginning as more trees are added, but the trend is downward and it stabilizes at about 40 trees. The patterns in the two error plots are similar. There is a bump up as the numbers of trees get in the range of 60-80 approximately, and then the error falls again and never goes up. Thus, increasing the number of trees beyond 80 or so does not really serve any purpose. Once we have seen the plot of errors, we can decide how many trees we need, if we want. Fewer trees will make testing faster. When building a random forest, the number of trees to build can be given as an argument called `ntree` to the `randomForest` command.

Line 32 of the program uses the `caret` library to print the confusion matrix and a number of metrics, like we have seen before.

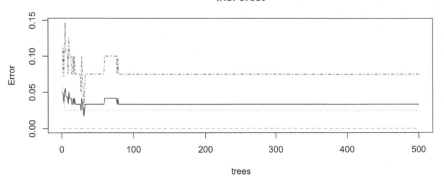

FIGURE 3.17: Misclassification error as trees are built in the random forest for the Iris dataset.

```
Confusion Matrix and Statistics

              Reference
Prediction   setosa versicolor virginica
  setosa        10          0         0
  versicolor     0          8         1
  virginica      0          2         9

Overall Statistics

               Accuracy : 0.9
                 95% CI : (0.7347, 0.9789)
    No Information Rate : 0.3333
    P-Value [Acc > NIR] : 1.665e-10

                  Kappa : 0.85
 Mcnemar's Test P-Value : NA

Statistics by Class:

                     Class: setosa Class: versicolor Class: virginica
Sensitivity                 1.0000            0.8000           0.9000
Specificity                 1.0000            0.9500           0.9000
Pos Pred Value              1.0000            0.8889           0.8182
Neg Pred Value              1.0000            0.9048           0.9474
Prevalence                  0.3333            0.3333           0.3333
Detection Rate              0.3333            0.2667           0.3000
Detection Prevalence        0.3333            0.3000           0.3667
Balanced Accuracy           1.0000            0.8750           0.9000
```

We can clearly see that the use of random forests brings up the values of metrics all around compared to using a single tree. In particular, all setosas are classified correctly, the versicolor classification remains the same as the single tree, and there is dramatic improvement in the virginica classification. This example clearly shows that the use of random forests has the potential to improve classification results dramatically. The iris dataset, although small, has 150 examples. Random forests are likely to work well when the number of examples is not very small. The two more toyish datasets we have used in this chapter, namely the Lenses (with 24 examples and 3 classes) and PlayTennis (with 14 examples and 2 classes), the use of random forests may not necessarily lead to better results, primarily because the test sets become real tiny with a handful of examples that can be counted on one's fingers.

When we have a single decision tree, it is easy to tell which variables are most important in classification and which ones are less important or not important at all. Variables or features that occur near the top of the tree are most important, and the importance goes down for variables that occur lower in the trees. Variables that do not occur in the tree are not at all important for classification purposes and can essentially be ignored. However, when we build a large number of decision trees with sampled training datasets and sampled features, it is difficult to tell which features are important and which ones are not. The `randomForest` library provides a way to gauge the importance of the features by counting how many times they occur in the built trees and how much error they reduce on average. This can be obtained using the `importance` command. For example, in the case of the `Iris` dataset, by running

```
importance (irisForest)
```

we get the following.

```
            MeanDecreaseGini
sLength           6.779879
sWidth            1.589782
pLength          35.346397
pWidth           35.657242
```

The importance of variables can also be plotted using the `varImpPlot()` command of the `randomForest` library. In this case, it produces the plot given in Figure 3.18. The listing and the plot show that across the 500 decision trees that have been built, `pLength` and `pWidth` are the most important features; they are almost equally important.

A random forest creates a large number of trees, by default 500 in this library. If we want to see the number of nodes in the trees produced, we can use the `treesize` command. In this particular case

```
1  treesize (irisForest)
```

produces the following.

```
  [1]   4   6   5   8   8   8   4   5   7   9   8   8   8   7   6   5   4   4   4   7   7  11   6   9   6   4  11   5
 [29]   5   6   4   9   8   3   4   4   5   8   7   9   4   7   6   6   9   4   7   8   6   6   3   7   4   5   5   7
 [57]   6   7   6   4  11   7   4   7   5   5   9   6   6   9   7   6   8   7   9   5   7   7   5   8   4   9   5  11
 [85]   5   5   7   7   5   5   7   8   8   6   8   4   7   5   9   8   6   8  12   7   3   7   5   7   6   6   4   7
[113]   6   6   8   7   6   9   6  11   3   5   7   4   7   7   4   4   4   4   6   5   6   8   8   7   8   7   6   8
[141]   3   6   4   5   7   9   9   4   9   4   8  12  10   9   4   9   5   5   6   5  11   6   6   9   6   9   6   9
[169]   5   4   5   9   7   8   5   6  11   8   4   4   6   5   5   5   7   6   6   8   6   5   6   5   6   8   4   5
[197]   7   8   7   8   4   7   5   5   4   4   7   6   5   8   8   7   9   4   8   6   7   7   4   4   4   7   8   3
[225]   8   4   7   6   9   6   5  10   6   7   4   4  11   5   6   9   7   9   4   6   4   5   5   8   5   5   6   4
[253]   4   4   7   4   7   6  10   6   4   7   4   6   6   7   6   5   4   9   6   5   4   7   6   5  10   6   7   8
```

```
[281]  8  4  4  7  7  4  6  9  5  8  9  7  7  8  5  8  6  4  4  6  4  5  7  5  6  6  6  6
[309] 10  7  5  5  3  6  7 11 10  4  8  7  8  4  5  7 10  7  4  9  5  9  9  9  7  6  5  8
[337]  6  8  4  7 12  6 10  6  4  9  8  4  8  3  4  6  4  9 10  4  9  5  6  5  7  3  5  7
[365] 11  4  3  9  6 13  9  8  6  6  7 10  7  7  7  5  6  5  7  5 12  7  4  7  6  4  8  3
[393]  5  7  6  5  8  5  8  8  8  4  7  4 10  5 10  8  5  4  4  9  4  7  7  6  9  8  9  5
[421]  7 10  7  7  6  4  7  3 10  7  4  7  5  4  4  7  8 10  6  5  6  5  6  4  8  6 10  8
[449] 10  7  7  8  3  5  6  6  4  8  6  7  7  5  5  5  6  4 11  7  6  6  7  5  8  4  3  8
[477]  7  6  5  8  7  7  5  5  8  5  4  5  4  6  4  4  8  5  7  8  9  8  9  7
```

For example, this output says that the 477th tree has a total of 7 internal nodes. We can reconstruct a specific tree by using the `getTree` command. For example

```
getTree (irisForest, k = 477)
```

produces

left daughter	right daughter	split var	split point	status	prediction
1	2	3	4	1.65	1	0
2	4	5	4	0.80	1	0
3	0	0	0	0.00	−1	3
4	0	0	0	0.00	−1	1
5	6	7	3	5.35	1	0
6	8	9	4	1.45	1	0
7	0	0	0	0.00	−1	3
8	0	0	0	0.00	−1	2
9	10	11	2	2.90	1	0
10	12	13	2	2.75	1	0
11	0	0	0	0.00	−1	2
12	0	0	0	0.00	−1	2
13	0	0	0	0.00	−1	3

Line 1 above corresponds to node 1 or the root node. It says that the left daughter is node 2 and right daughter is node 3. For splitting, it used variable number 4, the value used to split is 1.65. Status of 1 says it is a terminal node, −1 says it is an internal node. The prediction column says how many of the test examples were classified to be in that particular node. In this case,

irisForest

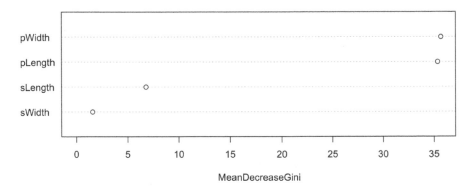

FIGURE 3.18: Variable importance in the random forest for the Iris dataset.

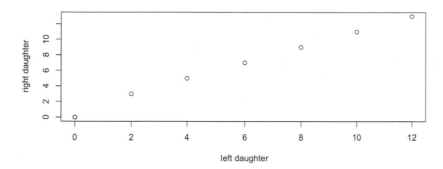

FIGURE 3.19: Graphical representation of a specific tree in a random forest.

we have a total of 16 test examples. Each tree is built to the largest size for classification by default, i.e., each tree is built unpruned by default. However, it is possible to provide an argument to the `randomForest` command to limit the size of trees built.

The number of features sampled is $\sqrt{p}$ where p is the total number of features, for decision trees. We can change the number of features sampled by giving a value to the parameter `mtry` in the `randomForest` command. In our case, we have only 4 features, so it samples two features at each split by default. We can plot a specific tree obtained by `getTree` of the `randomForest` library using the `plot` command provided by the library like the following.

```
1 plot (getTree (irisForest, k=477))
```

The plot is a more friendly representation of the table given above. The plot is given in Figure 3.19.

We can plot a specific tree like a normal tree also. The `randomForest` library does not have a tool to print a specific tree beyond what we have seen already. To do this, we need to install a library called `reprtree`, which is not a part of the standard R distribution.

```
1 library (devtools)
2 install_github('araastat/reprtree')
```

We should be now able to draw a specific tree using the following command.

```
1 t <- getTree (irisForest, k=477)
2 reprtree:::plot.tree(t, irisForest)
```

This produces the tree shown in Figure 3.20.

3.8 Boosting

Boosting is another way to put together an ensemble. Unlike random forests, the philosophy and approach are different. In building a random

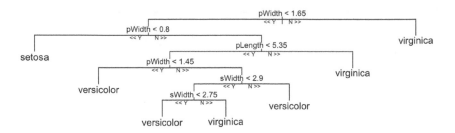

FIGURE 3.20: Graphical representation of tree number 477 in a random forest using the `reprtree` library.

forest, we generate a number of decision trees, which are built to the fullest. Each tree is built from a bootstrap dataset obtained by sampling with replacement from the original dataset, and by sampling a small number of features at every node. Although the ensemble contains only decision trees, each tree is sufficiently de-correlated from other trees, and the decisions are somewhat independent. Since a decision tree is a high-variance classifier to begin with, the semi-independent decision trees in a collection are likely to have varied opinions on the class of an unseen example $\vec{x}$. Each decision tree is an expert although fickle in nature. In summary, the random forest is a committee of experts that makes a decision together.

In contrast, boosting is guided by a different philosophy. Each classifier in a boosted ensemble is not an expert by any means, more of a simpleton. Unlike being semi-independent of each other, boosted participants are created in a sequence, where a participant in certain position in the sequence tries to better the collective decision of all that came before it. The first simple or "weak" classifier usually makes a decision based on one single attribute. We can illustrate a single-attribute classifier easily if we assume that the attribute is numerical. All examples are weighted equally for the first classifier. For example, a classifier can try to identify the type of iris flower by looking at petal length only, and splitting at a certain value of length. Because the classifier is simple, it is likely to make many mistakes in classifying unseen examples. In boosting, we expect it to be correct at least more than half the time, even if by a small margin. The second classifier is simple too, but has knowledge of errors the first classifier made. The second classifier knows on what examples the first classifier made mistakes and is asked to pay special attention to them, and at the same time worry a little less about the examples the first classifier got right. The second classifier takes this into account, and makes a classification decision, again using a single attribute. We can look at the mistakes made by the two classifiers together, and build another single-attribute classifier that wants to do better with the erroneous cases, paying a lower level of attention to the correctly classified cases. We build a sequence of K (a fairly large number, say 500) simple classifiers in this fashion. This is the essence of a boosted sequence of classifiers, where each simple classifier boosts the ability of the entire sequence so far by treating misclassified examples with more care while treating the correctly classified examples with a little less care as if to say they are easy ones to handle at this point. When a sequence of classifiers makes a cumulative decision, the decisions of individual classifiers are weighted differently, based on their individual performance.

3.8.1 Boosting on Loss: AdaBoost

Adaptive boosting or AdaBoost is an algorithm introduced by Freund and Schapire [10, 9]. It can be used to boost weak learners of various kinds, although single-node decision trees or decision stumps are commonly used.

AdaBoost is a binary classifier and requires that for each example $\langle x^{\vec{(i)}}, y^{(i)} \rangle$ in the dataset, the $y^{(i)}$ values are restricted to -1 or 1, standing for negative or positive classification, respectively. Adaboost can be initialized with a classifier $h^{(0)}()$ that just outputs 0 for all examples. It builds a sequence of classifiers, each adapted to the prior ones. Before the kth iteration, we have the sequence of classifiers $\mathcal{H}() = \langle h^{(0)}(), \cdots, h^{(k-1)}() \rangle$. The classification decision is made by a linear weighted combination of the $k - 1$ classifiers so far.

$$\mathcal{H}^{(k-1)}(\vec{x}) = \alpha_1 h^{(1)}(\vec{x}) + \cdots + \alpha_{k-1} h^{(k-1)}(\vec{x}). \tag{3.28}$$

Since $h^{(0)}()$ always outputs 0, it does not play a role in the collective classification, and is used simply to initialize before iteration. $\mathcal{H}^{(k-1)}()$ is the ensemble classifier so far. At the end of the kth iteration, we want to extend this to a better ensembled classifier by adding another weak classifier $h^{(k)}(\vec{x})$:

$$\mathcal{H}^{(k)}(\vec{x}) = \mathcal{H}^{(k-1)}(\vec{x}) + \alpha_k h^{(k)}(\vec{x}). \tag{3.29}$$

Our task is to identify the actual kth classifier $h^{(k)}()$ and its weight α_k.

Although AdaBoost was introduced by Freund and Schapire, we use a different derivation here, following [28]. AdaBoost uses an exponential error or loss function. The error for an individual classifier $h^{(k)}( )$ for an example $\langle \vec{x}, y \rangle$ is exponential: $e^{-y\, h^{(k)}(\vec{x})}$. In the exponent, we have a product of y and $h^{(k)}(\vec{x})$. We note that y, the actual label, is either -1 or 1. $h^{(k)}(\vec{x})$, which is the classification produced for $\vec{x}$ by the kth classifier, can be +ve, 0 or -ve. If the signs of y and $h^{(k)}(\vec{x})$ agree, we can say that the classification is correct resulting in a positive product, but a negative overall exponent. Thus, when the classifier classifies correctly, the error is negative power of e and hence small. If the sign of the two disagree, the classification is wrong and we have a positive exponent, resulting in a large error. If $h^{(k)}(\vec{x})$ comes out to be 0, the value of the expression is 1.

$L\big(\mathcal{H}^{(k)}(\vec{x}^{(i)})\big)$ is the total loss or error for a test example $\vec{x}^{(i)})$ for the ensemble being put together, now with k classifiers, obtained by weighted summing of the errors for the individual classifiers so far.

$$L\big(\mathcal{H}^{(k)}(\vec{x}^{(i)})\big) = e^{-y^{(i)}\,\mathcal{H}^{(k)}(\vec{x}^{(i)})}$$

To compute the loss over all N examples in the test dataset, we add the errors for all the examples.

$$L\big(\mathcal{H}^{(k)}\big) = \sum_{i=1}^{N} e^{-y^{(i)}\,\mathcal{H}^{(k)}(\vec{x}^{(i)})} \tag{3.30}$$

$$= \sum_{i=1}^{N} e^{-y^{(i)}\left[\mathcal{H}^{(k-1)}(\vec{x}^{(i)})+\alpha_k h^{(k)}(\vec{x}^{(i)})\right]}$$

$$= \sum_{i=1}^{N} e^{-y^{(i)}\,\mathcal{H}^{(k-1)}(\vec{x}^{(i)})} \times e^{-y^{(i)}\,\alpha_k h^{(k)}(\vec{x}(i))}$$

Assume $w^{(i)(k)}$ is the weight assigned to a data point $\vec{x}^{(i)}$ right before iteration k, and $w^{(i)(k)} = e^{-y^{(i)}\,\mathcal{H}^{(k-1)}(\vec{x}^{(i)})}$. Assume also that $w^{(i)(1)} = \frac{1}{N}$ for all $i = 1\cdots N$, i.e., each data point has a weight of $\frac{1}{N}$ to begin with. Assume also that after each iteration all weights are normalized so that they sum up to 1. We can now write

$$L\big(\mathcal{H}^{(k)}\big) = \sum_{i=1}^{N} w^{(i)(k)}\, e^{-y^{(i)}\,\alpha_k h^{(k)}(\vec{x}(i))}.$$

We can split this summation into two parts, ones for which $y^{(i)} h^{(k)}(\vec{x}(i)) = 1$ or the data examples that are correctly classified by the kth individual classifier, and ones for which $y^{(i)} h^{(k)}(\vec{x}(i)) = -1$ or the data examples are incorrectly classified by the kth individual classifier. Let us call these two subgroups *CorrectlyClassified* (CC) (=True Positives + True Negatives) and *Misclassified*(MC) (= False Positives + False Negatives).

$$L\big(\mathcal{H}^{(k)}\big) = \sum_{\vec{x}^{(i)}\in CC} w^{(i)(k)}\, e^{-y^{(i)}\,\alpha_k h^{(k)}(\vec{x}(i))} + \sum_{\vec{x}^{(i)}\in MC} w^{(i)(k)}\, e^{-y^{(i)}\,\alpha_k h^{(k)}(\vec{x}(i))}$$

$$= e^{-\alpha_k} \sum_{\vec{x}^{(i)}\in CC} w^{(i)(k)} + e^{\alpha_k} \sum_{\vec{x}^{(i)}\in MC} w^{(i)(k)}$$

$$= e^{-\alpha_k} \sum_{i=1}^{N} w^{(i)(k)} + \big(e^{\alpha_k} - e^{-\alpha_k}\big) \sum_{\vec{x}^{(i)}\in MC} w^{(i)(k)}$$

We include the terms for the $\vec{x}^{(i)} \in MC$ in the first sub-expression, and to keep the right-hand side the same as before, we subtract it in the second sub-expression. This means we need to search over attribute-value combinations to find the best attribute-value pair. This is likely to be a heuristic search as discussed in Section 3.3.2.

To choose the kth individual classifier $h^{(k)}(\vec{x})$ of the growing ensemble, we need to minimize the total error $L(\mathcal{H}^{(k)})$. Since the first sum adds the weights of elements over the entire set of N examples and does not change, to minimize $L(\mathcal{H}^{(k)})$, we need to minimize the second sum $(e^{\alpha_k} - e^{-\alpha_k}) \sum_{\vec{x}^{(i)} \in MC} w^{(i)(k)}$. In this summation, $(e^{\alpha_k} - e^{-\alpha_k})$ is constant for a specific k, and therefore, we actually need to minimize $\sum_{\vec{x}^{(i)} \in MC} w^{(i)(k)}$, which is the sum of the weights of the misclassified individual data points in the kth iteration. So, to choose the kth classifier, we need to search for a decision stump (i.e., a single attribute and a corresponding value for the attribute) that minimizes the weights of the misclassified examples.

Once we have chosen the classifier, to determine the weight α_k for the kth classifier, we differentiate $L(\mathcal{H}^{(k)})$ with respect to α_k and set it to 0.

$$\frac{d}{d\alpha_k} L(\mathcal{H}^{(k)}) = \frac{d}{d\alpha_k} \left[e^{-\alpha_k} \sum_{\vec{x}^{(i)} \in CC} w^{(i)(k)} + e^{\alpha_k} \sum_{\vec{x}^{(i)} \in MC} w^{(i)(k)} \right]$$

$$= -e^{-\alpha_k} \sum_{\vec{x}^{(i)} \in CC} w^{(i)(k)} + e^{\alpha_k} \sum_{\vec{x}^{(i)} \in MC} w^{(i)(k)}$$

$$= 0$$

We solve to find α_k.

$$e^{-\alpha_k} \sum_{\vec{x}^{(i)} \in CC} w^{(i)(k)} = e^{\alpha_k} \sum_{\vec{x}^{(i)} \in MC} w^{(i)(k)}$$

$$\sum_{\vec{x}^{(i)} \in CCs} w^{(i)(k)} = \left(e^{\alpha_k}\right)^2 \sum_{\vec{x}^{(i)} \in MC} w^{(i)(k)}$$

Continuing with the solution, we get the following.

$$\left(e^{\alpha_k}\right)^2 = \frac{\sum_{\vec{x}^{(i)} \in CC} w^{(i)(k)}}{\sum_{\vec{x}^{(i)} \in MC} w^{(i)(k)}}$$

$$e^{\alpha_k} = \sqrt{\frac{\sum_{\vec{x}^{(i)} \in CC} w^{(i)(k)}}{\sum_{\vec{x}^{(i)} \in MC} w^{(i)(k)}}}$$

Finally, we get

$$\alpha_k = \frac{1}{2} \ln \left(\frac{\sum_{\vec{x}^{(i)} \in CC} w^{(i)(k)}}{\sum_{\vec{x}^{(i)} \in MC} w^{(i)(k)}} \right)$$

$$= \frac{1}{2} \ln \left(\frac{\sum_{i=1}^{N} w^{(i)(k)} - \sum_{\vec{x}^{(i)} \in MC} w^{(i)(k)}}{\sum_{\vec{x}^{(i)} \in MC} w^{(i)(k)}} \right)$$

$$= \frac{1}{2} \ln \left(\frac{1 - e^{(k)}}{e^{(k)}} \right)$$

where $e^{(k)}$ is the ratio of the sum of the weights for the misclassified examples to the sum of weights of all the examples in the kth iteration. This assumes that at the end of each iteration, the weights of the elements are normalized so that they sum up to 1.

Thus, we have an algorithm at this point in time for AdaBoost. The algorithm initializes by assigning each element a weight of $\frac{1}{N}$ to start. It goes through a loop where in each iteration, it does the following four things. (1) It builds a weak classifier (usually, a stump using just single attribute) that minimizes the weighted misclassification error. (2) It assigns a new weight to each element, accentuating the weights on the misclassified examples and at the same time reducing the weights on the correctly classified examples. (3) It makes sure that the sum of the weights of all the elements is 1 by normalizing. (4) It also computes a weight for the newly introduced weak classifier. It executes the loop a large number of times, which could be as large as a few hundred. The AdaBoost algorithm is given as Algorithm 3.1. To test the ensemble on a data example $\vec{x}$, the program simply outputs $sign\left(\sum_{k=1}^{K} \alpha^{(k)} h^{(k)}(\vec{x})\right)$.

Algorithm 3.1: Algorithm for AdaBoost

Input: Training Set $\mathbf{Z} = \langle \vec{x}^{(i)}, y^{(i)} \rangle, i = 1 \cdots N$

Output: A set of classifiers $h^{(1)} \cdots h^{(K)}$ with corresponding weights $\alpha^{(k)}$

/* Initialize weights of individual examples   */

1 $w^{(i)(1)} \leftarrow \frac{1}{N}$ for $i = 1 \cdots N$;

/* K is the total number of trees to build   */

2 **for** $k = 1$ to K **do**

    /* Compute weight for kth classifier   */

3      $h^{(k)}(\ ) \leftarrow$ a weak classifier that minimizes misclassification error

        $\sum_{\vec{x}^{(i)} \in MC} w^{(i)(k)}$

    /* $e^{(k)}$ is the ratio of sum of weights for misclassified
       examples to sum of weights of all examples in kth
       iteration   */

4      $\alpha^{(k)} \leftarrow \frac{1}{2} ln\left(\frac{1-e^{(k)}}{e^{(k)}} \right)$;

    /* Update weights for data examples   */

5   **for** $i=1$ to N **do**

6   **if** $\vec{x}^{(i)}$ *is misclassified by* $h^{(k)}$$) < 0$ **then**

            /* i.e., it is a miss; make weight bigger   */

7              $w^{(i)(k+1)} \rightarrow w^{(i)(k)} \sqrt{\frac{1-e^{(k)}}{e^{(k)}}}$;

8   **else**

            /* make weight smaller   */;

9              $w^{(i)(k+1)} \rightarrow w^{(i)(k)} \sqrt{\frac{e^{(k)}}{1-e^{(k)}}}$;

10   Normalize all $w^{(i)(k+1)}$s to add up to 1;

3.8.1.1 Multi-Class AdaBoost

AdaBoost, as presented in the previous subsection, is a binary classifier. To perform multi-class classification, as in the case of the Iris classification problem we have often discussed in this chapter, it needs to be converted to multiple binary classification problems. There are usually two approaches to do this: one-against-all and one-against-one. Additionally, it is possible to create new versions of the binary classifier that are truly multi-class to begin with. However, this often requires redeveloping the mathematics behind the classifier.

One-against-all classification: Suppose we have m classes: $C_1, \cdots, C_m$. We can create m derived classification problems from the original problem. Each problem is formulated as classifying examples of class C_i from examples of all other classes, i varying from 1 to m. This requires creating derived m datasets from the original dataset.

For example, to classify class C_1 against all, create a new dataset in which every example from classes $C_2, \cdots, C_m$ is relabeled as a new class, say C_{1-}. Now, we build a classifier for this new classification problem. This new classifier, after training, should return a class label as well as a confidence label or probability when it predicts the class of an unseen example. So, for a new example $\vec{x}$, this new classifier tells us the probability that it belongs to class C_1. The unknown example is classified by each one of the newly created m classifiers. To obtain the class $\vec{x}$ belongs to, we pick the class C_i that gets the most confidence or probability.

Since in an individual dataset we create for class C_i, all examples that do not belong to the class are lumped together, the dataset is unlikely to be balanced. Many classifiers have problems when they are trained on unbalanced classes. However, decision trees do not have this drawback.

One-against-one classification: In this approach also, we create a number of classifiers, but in a different way. For example, to classify if an unseen example $\vec{x}$ belongs to class C_1, we train m classifiers: C_1 vs. C_2, C_1 vs. C_3, and so on. Thus, to train these classifiers, we create derived datasets, picking examples from only the relevant classes. Thus, in this case, we create $\frac{1}{2}m(m+1)$ classifiers as shown below.

C_1 vs. C_2, C_1 vs. C_3, C_1 vs. $C2, \cdots, C_1$ vs. C_m
C_2 vs. C_3, C_2 vs. $C_4, \cdots, C_2$ vs. C_m
$\vdots$
C_{m-1} vs. C_m

If we look carefully, we see that each class C_i participates in m direct competitions with another class $C_{j, j \neq i}$. When we want to classify an unknown example $\vec{x}$, each of these classifiers vote. Each class can get up to m win votes. The class that gets the most votes wins. This approach creates a large number of classifiers and becomes time-consuming.

Multi-class AdaBoost Algorithm SAMME: SAMME (Stagewise Additive Modeling for Multi-class Exponential loss function) is an extension of the AdaBoost Algorithm to multi-class classification [2]. The SAMME algorithm can be written out like the AdaBoost algorithm only with one difference. The constant $\alpha^{(k)}$ is computed a bit differently, taking into account the increased number of classes. The AdaBoost algorithm deals with two classes and requires that a weak classifier at any stage has an accuracy which is greater than or equal to $\frac{1}{2}$. SAMME works with m classes and simply requires that the weak classifier be better than random guess, which in this case, is an accuracy of $\frac{1}{m}$ instead of $\frac{1}{2}$. Thus, if we have 10 classes, each weak classifier has to perform better than only 10%. As a result, line 4 of the AdaBoost algorithm will change to

$$\alpha^{(k)} \leftarrow ln\left(\frac{1 - e^{(k)}}{e^{(k)}}\right) + ln(m - 1). \tag{3.31}$$

The derivation can be found in [2]. Given this algorithm, we can write that

$$\alpha^{(k)} \leftarrow ln\left(\frac{1 - e^{(k)}}{e^{(k)}}(m - 1)\right)$$

and the need for $\alpha^{(k)}$ to be positive gives us the inequality given below, which has been simplified further.

$$\frac{1 - e^{(k)}}{e^{(k)}}(m - 1) > 1 \tag{3.32}$$

$$\frac{1 - e^{(k)}}{e^{(k)}} > \frac{1}{m - 1}$$

$$\frac{1}{e^{(k)}} - 1 > \frac{1}{m - 1}$$

$$\frac{1}{e^{(k)}} > 1 + \frac{1}{m - 1}$$

$$\frac{1}{e^{(k)}} > \frac{m - 1}{m}$$

$$e^{(k)} < 1 - \frac{1}{m}$$

$$1 - e^{(k)} > \frac{1}{m}$$

The last inequality tells us that $1 - e^{(k)}$, which is the error of the kth classifier for $k = 1 \cdots K$, must be bigger than $\frac{1}{m}$, as discussed above.

3.8.1.2   AdaBoost in R

There are several packages that implement AdaBoost in R, for binary as well multi-class cases. In this book, we present the `adabag` package that implements the SAMME algorithm, which works for both binary and multi-class classification. In the program below, we work with the Iris dataset that has three class labels, and hence, we use the SAMME algorithm, although it has an option for using the AdaBoost.M1 algorithm also.

```
1   #file: irisAdaBoostSAMME.r
2   library (caret) #for confusionMatrix
3   library (adabag)   #to build random forest
4   library (readr) #for read_csv
5
6   iris <- readr::read_csv("Datasets/txt/iris.txt")
7   View(iris)
8
9   #R reads the column values as strings; convert the class column into factor,
10  #i.e., enumerated values
11  iris$iType <- as.factor (iris$iType)
12  #read_csv reads as a tibble, which needs to be converted to a
13  #dataframe for adabag to work
14  iris <- as.data.frame (iris)
15
16  set.seed(2)
17  #80% training and 20% testing
18  trainIndex <- caret::createDataPartition (y=iris$iType, p=0.8, list = FALSE)
19  irisTrainingSet <- iris [trainIndex,] #obtain the training set
20  irisTestSet <- iris [-trainIndex,]     #obtain the test set
21
22  #Build a sequence of boosted trees trained on the training set
23  #data should be a dataframe but it doesn't work belowirris
24  #Zhu says use SAMME as the algorithm
25  #Building only stumps using rpart control
26  irisAdaTrees<- adabag::boosting (iType ~ ., data = irisTrainingSet, boos=TRUE,
27                  mfinal=100, coeflearn = 'Zhu', control=rpart.control(maxdepth=1))
28
29  summary (irisAdaTrees)
30
31  irisAdaTrees$weights
32  irisAdaTrees$votes
33  irisAdaTrees$importance
34
35  #plotting a specific tree
36  plot (irisAdaTrees$trees [[50]])
37  text (irisAdaTrees$trees [[50]], pretty=0)
38
39  irisAdaMargins <- margins (irisAdaTrees, iris)
40  plot.margins (irisAdaMargins)
41
42  adabag::errorevol(irisAdaTrees, iris)
43  plot.errorevol(adabag::errorevol(irisAdaTrees, iris))
44
45  #obtain predictions for the test set
46  #Gives probabilities for the three classes: setosa, versicolor, virginica
47  irisPredictions <- predict (irisAdaTrees, newdata = irisTestSet)
48  #actual labels for the classes in irisTestSet
49  irisTestLabels <- irisTestSet$iType
50
51  #prints out confusion matrix and other details
52  caret::confusionMatrix(irisPredictions$class, irisTestLabels)
```

As usual, lines 1 through 5 load the libraries needed, and load the dataset and view the dataset. Line 11 converts the class label to a factor or discrete value. Line 14 converts the data read into a dataframe as needed by the adabag library. Lines 16 through 20 use a random number to divide the dataset into training and test data subsets, with 80% in the training set. Line 26 repeated below, builds 100 boosted stumps using the SAMME algorithm.

```
irisAdaTrees<- adabag::boosting (iType ~ ., data = irisTrainingSet, boos=TRUE,
              mfinal=100, coeflearn = 'Zhu', control=rpart.control(maxdepth=1))
```

We use the `adabag::boosting` function to create the boosted trees by stating `boos=TRUE` . The formula to create the trees is `iType ~ .`, indicating that all features should be used. Here, we use the training set to create the trees. The number of trees grown is 100, but for bigger datasets, one may grow more trees, say 500. `coeflearn = 'Zhu'` says that we should use the SAMME algorithm; Zhu was its inventor. We want to build stumps as the individual trees. This is achieved by using what is called the `rpart.control`, which in this case states that the maximum depth of the trees is 1. Other depths are definitely possible. The process of building the trees takes a while as can be noticed.

Once the sequence of boosted trees are built, we can obtain a summary of what was built using the **summary** command. The boosted trees built are stored in an object, here called `irisAdaTrees`. We can obtain the weights given to the 100 trees built in this case by writing the following.

```
irisAdaTrees$weights
```

This produces the following, which shows the weights assigned to the votes of each of the hundred trees.

```
 [1] 1.3862944 2.3025851 2.7838883 1.7837552 2.4013340 2.8139676 2.9129129 2.2762943 1.8810902 1.9554702
[11] 2.5033773 1.7465018 2.3292882 1.2532298 2.5163876 1.8188710 1.7402365 1.4009474 2.3077885 1.7059279
[21] 1.7043856 1.5528112 2.5910383 1.6976541 1.5554571 1.5338601 1.4593507 1.6676108 1.6187129 1.9636159
[31] 1.6149043 2.0883112 1.9158671 2.0039243 1.3848516 1.7440858 1.9379659 1.6069579 1.8750248 1.5974063
[41] 1.6690676 2.1237501 1.6606931 1.6816256 2.3061640 1.7842261 1.9756552 1.7461156 1.3688157 2.0589462
[51] 1.5523644 1.5996230 2.2219580 1.8522370 2.1410658 1.6652701 1.7044029 2.0177051 1.1893321 1.4152541
[61] 1.7943919 1.7266804 1.6428452 1.5591813 1.9928152 1.6237376 2.2197585 1.6852112 2.1591894 1.6699864
[71] 1.9564454 1.8332446 2.0304959 1.6110086 1.8449700 1.8565448 1.5974691 1.2905857 2.1015684 1.8887037
[81] 1.5911925 1.7553404 1.7171872 1.6893701 1.6421813 1.8764881 2.2207625 1.3912144 1.8624300 1.5274904
[91] 1.9282465 1.5056286 1.5449555 1.2994237 1.9731838 2.2212477 1.6855753 1.7342063 1.7029238 0.9374047
```

We can use `irisAdaTrees$votes` to see how each one of the 100 trees voted for the three classes of iris for each example. Below, we see the votes of the first 20 trees.

```
 [1,]  103.895807   63.06144   15.602258
 [2,]  102.436456   59.78879   20.334253
 [3,]  103.895807   63.06144   15.602258
 [4,]  103.895807   63.06144   15.602258
 [5,]  103.895807   63.06144   15.602258
 [6,]  103.895807   64.93646   13.727233
 [7,]  103.895807   63.06144   15.602258
 [8,]  103.895807   63.06144   15.602258
 [9,]  102.436456   59.78879   20.334253
[10,]  103.895807   63.06144   15.602258
[11,]  103.895807   63.06144   15.602258
[12,]  102.436456   59.78879   20.334253
[13,]  102.161601   70.45629    9.941607
[14,]  103.895807   64.93646   13.727233
[15,]  103.895807   63.06144   15.602258
```

```
[16,] 103.895807   64.93646   13.727233
[17,] 103.895807   63.06144   15.602258
[18,] 103.895807   63.06144   15.602258
[19,] 103.895807   63.06144   15.602258
[20,] 103.895807   63.06144   15.602258
```

We can obtain the importance of the variables by using `irisAdaTrees$`
`importance`. We can plot a specific tree also, as we do in lines 36-37.

We have discussed earlier that the concept of margins is important in
building boosted trees using AdaBoost. We can plot the frequencies of the
various margin values that occur in the 100 trees produced. Such a plot is seen
in Figure 3.21. It shows that the margins are high, indicating the classification
is good. We can plot the evolution of the errors as the number of trees increases
using the command

```
plot.errorevol(adabag::errorevol(irisAdaTrees, iris))
```

Line 42 prints the evolution of the error of classification as the number of trees
increases.

```
  [1] 0.33333333 0.33333333 0.04666667 0.04666667 0.04666667 0.04666667 0.04000000 0.04666667 0.02666667
 [10] 0.02666667 0.02000000 0.02000000 0.04000000 0.03333333 0.02000000 0.02666667 0.02000000 0.02666667
 [19] 0.02000000 0.02000000 0.02666667 0.02000000 0.02000000 0.02000000 0.02000000 0.02000000 0.02000000
 [28] 0.02000000 0.02000000 0.02000000 0.02000000 0.02000000 0.02000000 0.02000000 0.02000000 0.02000000
 [37] 0.02000000 0.02000000 0.02000000 0.02000000 0.02000000 0.02000000 0.02000000 0.02000000 0.02000000
 [46] 0.02000000 0.02000000 0.02000000 0.02000000 0.02000000 0.02000000 0.02000000 0.02000000 0.02000000
 [55] 0.02000000 0.02000000 0.02000000 0.02000000 0.02000000 0.02000000 0.02000000 0.02000000 0.02000000
 [64] 0.02000000 0.02000000 0.02000000 0.02000000 0.02000000 0.02000000 0.02000000 0.02000000 0.02000000
 [73] 0.02000000 0.02000000 0.02000000 0.02000000 0.02000000 0.02000000 0.02000000 0.02000000 0.02000000
 [82] 0.02000000 0.02000000 0.02000000 0.02000000 0.02000000 0.02000000 0.02000000 0.02000000 0.02000000
 [91] 0.02000000 0.02000000 0.02000000 0.02000000 0.02000000 0.02000000 0.02000000 0.02000000 0.02000000
[100] 0.02000000
```

Figure 3.22 shows how the error evolves as the number of trees becomes larger.
It clearly shows that the largest drop in error happens within the first 5 trees
or so, and the error does not drop after 20 trees or so, and hence, it is not
necessary to build a 100 trees. To test how the trained sequence of boosted

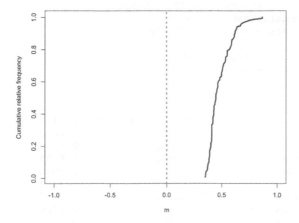

FIGURE 3.21: Plot of margin frequencies over 100 trees.

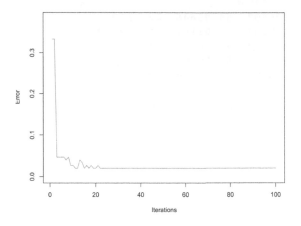

FIGURE 3.22: Plot of error evolution over 100 trees.

trees classify, we use the test set created in line 20. The `predict` statement is used to predict classification labels for the test set in line 47. We create a confusion matrix in line 52, which also provides additional details as given below.

```
Confusion Matrix and Statistics

              Reference
Prediction    setosa versicolor virginica
  setosa          10          0         0
  versicolor       0          8         1
  virginica        0          2         9

Overall Statistics

               Accuracy : 0.9
                 95% CI : (0.7347, 0.9789)
    No Information Rate : 0.3333
    P-Value [Acc > NIR] : 1.665e-10

                  Kappa : 0.85
 Mcnemar's Test P-Value : NA

Statistics by Class:

                     Class: setosa Class: versicolor Class: virginica
Sensitivity                 1.0000            0.8000           0.9000
Specificity                 1.0000            0.9500           0.9000
Pos Pred Value              1.0000            0.8889           0.8182
Neg Pred Value              1.0000            0.9048           0.9474
Prevalence                  0.3333            0.3333           0.3333
Detection Rate              0.3333            0.2667           0.3000
Detection Prevalence        0.3333            0.3000           0.3667
Balanced Accuracy           1.0000            0.8750           0.9000
```

3.8.2　Regression Using Trees

In this section, we will introduce the topic of regression using trees. We can perform regression using individual trees, or an ensemble of trees such as random forests and loss-boosted trees we have discussed in this chapter so far.

3.8.2.1　Regression Using Single Trees

We have discussed at length in Section 3.3 how a single decision tree can be built for classification from training data. The only difference between classification and regression trees is that the label or the dependent variable in classification is a discrete value whereas in regression, it is numeric or a real number. In Chapter 2, we have discussed regression in depth; the methods discussed in this chapter, produce algebraic formulas that minimize an error or loss function, possibly with regularization.

In this subsection, we show that a decision tree can be easily adapted to perform regression. The only thing that needs to change is how we decide how to find a feature to split on and the value to use for the feature to split. The dependent variable in regression is numeric. So, we cannot use a metric such as entropy or Gini coefficient to decide on a splitting feature and its value, and we use variance instead. As we know, there are two cases. First, the candidate splitting variable is discrete, with a small number of possible values, and second, the candidate splitting variable is numeric with real values. We have had a similar discussion in Subsection 3.3.2, but repeat here for regression.

Let the candidate splitting variable j_j be discrete in the dataset D. Instead of computing $\Delta Gini(D, f_j)$ or $\Delta Entropy(D, f_j)$, we use variance. The values of variable f_j divide the dataset D into several subsets. For each subset, we compute the variance in the dependent variable. We also compute the variance of the entire set D before splitting. Then, like the change in entropy or Gini coefficient, we can compute the change in variance between before and after:

$$\Delta Var(D, f_j) = Var(D) - \sum_{i=1}^{nv(f_j)} \frac{|D_{f_i}|}{|D|} Var(D_{f_i}) \qquad (3.33)$$

where D is the original dataset, $nv(f_j)$ is the number of distinct values feature f_j takes, $|D|$ is the number of examples in dataset D, and $|D_{fi}|$ is the number of total examples in the ith branch of the tree, i.e., the number of examples in D that have the ith unique value of f.

Consider the case where the candidate splitting variable is numeric. We can use a heuristic to find a number of candidate split points. For example, if the number of points is small, we can propose a split between every two consecutive points, with each splitting condition being of the form $f_j \leq value$. If the number of points is large, we can divide the range of values into quartiles (4 parts) or deciles (10 parts) or even more, and propose a number of split points. Each split is binary. At each candidate split point, we compute

$\Delta Var(D, f_j)$, considering the variance of the dependent variable before split and the variances on the two sides after split.

Thus, given all the features in the examples in the dataset D, we enumerate all the potential split points, and then compute the change in variance for all potential split points, and pick the one that leads to the largest change in variance.

3.8.2.2 Regression Using Random Forests

For regression using random forests, we build k number of trees as usual by sampling the dataset with replacement, and by sampling the features at each split point. For splitting, we use the change in variance measure. Given a test data example, each tree in the random forest makes a prediction regarding the value of the dependent variable. The values produced by the k trees are averaged to make the actual prediction.

3.8.2.3 Regression Using Loss-Boosted AdaBoost Trees

The AdaBoost SAMME algorithm, as implemented in the **adabag** library can perform classification only. It cannot be used for regression.

3.8.2.4 Regression in R Using Trees

We can perform regression using the **tree** library we have discussed in Section 3.4. The program given below builds such a regression tree, using the dataset called *Immigrants*, we first introduced in the beginning of Chapter 2.

```
1  #Create a regression tree with the tree library and plot it
2  library (tree)
3
4  Immigrants <- read.csv("~/Datasets/txt/Immigrants.csv")
5  View (Immigrants)
6  plot (Immigrants)
7
8  immigrantTree <- tree(Immigrants$Income ~ ., data = Immigrants)
9
10 plot (tree1)
11 text (tree1, pretty = 0)
```

The program is more or less the same as the first program in Section 3.4 that constructs a classification tree for the **lenses** dataset. It reads the **Immigrants** dataset and builds a regression tree for the dependent variable called **Income**. Note that a regression tree can use the same feature again and again, unlike a decision tree. The regression tree built is shown in Figure 3.23.

We can use random forests to perform regression as well. The following program creates 500 regression trees to predict the value of **Income** for the same dataset. The error plot as the number of trees increases is given in Figure 3.24. From the plot, it is clear that there is no need to build beyond 10 or so nodes to minimize the total error.

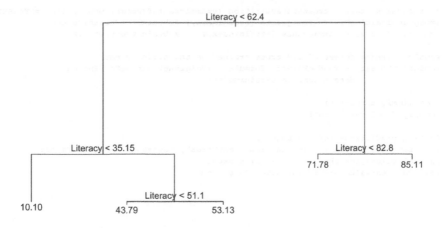

FIGURE 3.23: A simple regression tree for the Immigrants dataset.

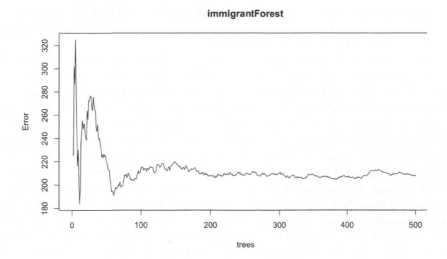

FIGURE 3.24: The error plot of the random forest for the Immigrant dataset as the number of trees increases.

```
1   #Regression tree with caret, training and testing sets
2   library (caret) #for confusionMatrix data splitting
3   library (randomForest)   #to build random forest
4
5   Immigrants <- read.csv("~/Datasets/txt/Immigrants.csv")
6   View (Immigrants)
7   plot (Immigrants)
8
9   #set a seed to generate random numbers
10  set.seed(2)
11  #80% training and 20% testing
```

```
12  trainIndex <- caret::createDataPartition (y=Immigrants$Income, p=0.8, list = FALSE)
13  immigrantTrainingSet <- Immigrants [trainIndex,] #obtain the training set
14  immigrantTestSet <- Immigrants [-trainIndex,]     #obtain the test set
15
16  #Build a random forest of 500 trees trained on the training set
17  immigrantForest <- randomForest (formula = immigrantTrainingSet$Income ~ .,
18                  data = immigrantTrainingSet)
19
20  plot (immigrantForest)
21  summary (immigrantForest)
22
23  #obtain predictions for the test set
24  immigrantPredictions <- predict (immigrantForest, newdata = immigrantTestSet)
25  #actual values for the classes in irisTestSet
26  immigrantTestValues <- immigrantTestSet$Income
```

3.9   Summary

In this chapter, we have discussed a number of ways to build decision trees. Although the focus has been mostly on building classification trees, the same ideas can be used for building regression trees.

We started with discussions on building single decision trees, using Gini index or entropy as splitting criterion for nodes. Although single decision trees are easy to build and interpret, they are not very good classifiers. They are unstable and high variance classifiers in the sense that small changes in the training dataset may result in completely different trees that may perform differently on test examples. To make the use of decision trees robust and dependable, ensembles of decision trees are widely used. Ensembles can be constructed in several ways. Random Forests are a collection of a large number of straightforward decision trees, each built to the fullest, using the same training dataset but by sampling training examples with replacement as well as by sampling a small number of features. The decisions of individual trees are combined (e.g., by majority voting) to obtain the final classification for a test example. Random Forests are robust and efficient classifiers that are used widely in various application domains. Another fruitful way to build an ensemble of decision trees is to build a sequence of "weak" classifiers. A boosted sequence of decision trees also makes for a strong classifier. Although the theoretical discussions are beyond the scope of this boo, a boosted algorithm called XGBOOST [6] has been very successful in classification tasks in the recent past.

Exercises

1. Suppose we have a dataset with N examples with binary labels. Assume also that there are f features in each example. For simplicity, assume each feature is binary, taking two values. How many distinct decision trees can be drawn with such a dataset? What if each feature can take up to n distinct discreet values, and there are l possible labels.

2. Decision trees are built in a greedy manner because the number of possible trees that can be drawn is very large. What is the time complexity of a decision tree building algorithm if the dataset has N example, each example with f features—with each feature taking up to n discreet distinct values.

3. Decision trees are built using greedy algorithms. As a result, a tree built is likely to be sub-optimal and find a local optimum in the space of possible trees. How can one improve the decision produced by such a sub-optimal classification model?

4. We discussed regularization in the context of linear least squared regression. Can the idea of regression be applied to decision trees? Discuss how you think regularization can be applied to decision tree building.

5. Missing values in data examples causes problems for any machine learning algorithm. Can we simply ignore examples with missing values? When is it appropriate to do so? If data examples with missing values cannot be ignored, what an be done to use the examples with missing values? What are some general approaches to do so?

6. Just like classification, regression can be performed in many different ways. How do you compare regression discussed in the previous chapter (least squared linear regression with regularized versions) and regression discussed in this chapter by using decision trees, and ensembles of decision trees. What are some other ways to perform regression?

7. In building random forests, data examples are picked with replacement. What is the reason for replacement? What differences are likely to come about in classification results due to the use or non-use of replacement?

8. XGBOOST is an algorithm that builds a sequence of regularized boosted trees. How does XGBOOST incorporate regularization in the optimization process?

9. Deep learning or artificial neural networks are the most productive or talked-about machine larning algorithms at this time. However, XG-BOOST is also highly regarded and produces excellent results for certain

problems. What kind of problems are likely to obtain excellent results using XGBOOST?

10. In this chapter, we discussed the building of an ensemble of tree classifiers in the form of random forests. Can ensembles be made from classifiers that are not the same type? What are some considerations in building such an ensemble of classifiers?

Chapter 4

Artificial Neural Networks

In this chapter, we discuss Artificial Neural Networks (ANNs), an approach to machine learning that has been around more than fifty years, but has gained great prominence since the early 2010s. Early ANNs usually had two or three layers of neuronal elements. Recently, the number of layers has increased substantially to tens of layers or even hundreds. Training of neural networks has also become more efficient and effective. In addition, the sizes of datasets that can be used for training have also increased substantially. Very large amounts of data are being generated by commercial and non-profit organizations and governments. Many of these datasets are available for training machine learning algorithms, including deep ANNs. The use of deep ANNs is called Deep Learning.

4.1 Biological Inspiration

Most of us believe that human beings are the most evolved living forms in the history of the universe, endowed with superior intelligence, with abilities for sophisticated reasoning, use of complex language, and introspection in addition to the felicity with use of fire, and manipulation of things with fingers and tools.

Superior human mental faculties and physical dexterity are a by-product of a highly developed brain. Like any other organ in the body, the human brain is composed of cells. The predominant type of cells in the brain are the neurons, and scientists estimate that there are about ninety billion of them. The neurons are highly inter-connected, with an average neuron being connected to 7,000 other neurons. A neuron has its characteristic shape. The cell body, which includes the nucleus, is the part of the neuron that controls the activities of the neuron. The cell body has protrusions that are called dendrites, which branch out to what can be called dendritic fibers. The cell body, also has something that looks like a long tail called an axon. An axon can be as long as a meter or more in length, and branches out to axonal fibers at the end.

A neuron or a nerve cell communicates with other neurons via connections called synapses that connect tips of dendritic branches with tips of axonal

DOI: 10.1201/9781003002611-4

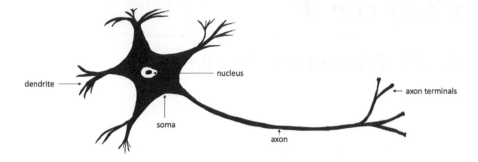

FIGURE 4.1: A Single Neuron

fibers of preceding neurons. A neuron is electro-chemically excitable. In other words, a neuron receives electrico-chemical signals as inputs from preceding neurons via the synapses, and integrates all the inputs it receives in the nucleus, and if the totality of the inputs exceeds a certain threshold, the neuron gets activated or it fires, and sends its output along the axon and then the synapses at the axonal fibers.

In particular, consider two important functions: the interpretation of what we see and what we hear. The original signals, in the case of vision, come from the visual organs, which are the eyes, each of which has links to a large number of neurons behind them. These nerves receive input from the camera-like machinery of the eyes. These neuronal collections or nerve fibers are grouped together like a cable, suitably called the optic cable, which connects and transmits visual information to the visual cortex of the brain. A cortex is a signal processing region of the brain that sits somewhat in its periphery. The visual cortex converts the received visual signals through neural processing to a perception or knowledge of the objects that we see. By neural processing, we refer to signals moving from one neuron to another to another and so on, and ultimately turning to actual information or useful knowledge.

In the case of speech, the auditory nerve connects the inner ear directly to the auditory cortex spanning both sides of the brain, where sound is processed in a similar manner. The further processing of visual or auditory signals that travel to the visual or auditory cortex takes place through a network or specialized neurons located in these areas of the brain. These regions, like other regions of the brain are made up of a complex network of neurons, one connected to the other. As the signals pass through a network of neurons, interpretation and understanding take place, in the way we commonly understand. We discuss these two examples cursorily only, to motivate why scientists and engineers have been inspired by the brain circuitry in the quest to develop computational devices and techniques to process vision and speech related problems, as well as other problems in artificial intelligence.

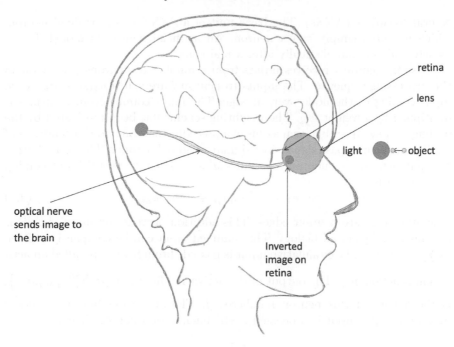

FIGURE 4.2: Visual Processing

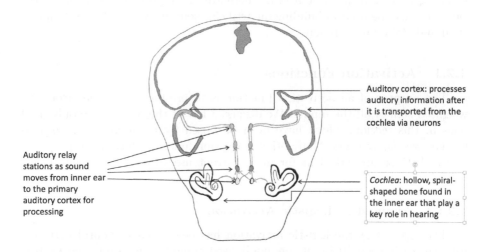

FIGURE 4.3: Auditory Processing (Adapted from www.mayoclinic.org)

4.2 An Artificial Neuron

To perform computing by mimicking how processing happens in the brain, computer scientists and engineers have developed what are called Artificial

Neural Networks (ANNs). The basic unit of an ANN is an artificial neuron, which we often simply call a neuron. An artificial neuron is modeled very loosely and very simplistically after a real neuron.

Assume neuron i receives inputs from a number of neurons prior to it in the processing sequence. The inputs to neuron i are called $a_{1i}, \cdots, a_{ni}$, each input coming to the node over an edge. The input coming from neuron j is multiplied by a weight w_{ji}. The weighting scheme has been motivated by the finding that synapses between axons of a preceding neuron and the dendrites of a following neuron pass signals in a similar weighted manner. The contribution of input neuron j to neuron i becomes $w_{ji}a_{ji}$. Thus, the total input coming to neuron i is

$$w_{1i}a_{1i} + \cdots w_{ji}a_{ji} + \cdots w_{ni}a_{ni} \tag{4.1}$$

assuming there are n input edges. This addition of weighted input signals is also inspired by actual biology. This summed input can be compactly written as $\sum_{j=1}^{n} w_{ji}a_{ji}$. The cumulative input is passed through what is called an activation function $g(\ )$. The output of the activation function, $g\left(\sum_{j=1}^{n} w_{ji}a_{ji}\right)$, is the output of this neuron numbered j. The activation function crudely mimics how the input is processed in the nucleus of a neuron. Thus,

$$o_i = g\left(\sum_{j=1}^{n} w_{ji}a_{ji}\right), \tag{4.2}$$

where o_i is the output of neuron i. There are many possible activation functions that can be used in building an artificial neuron. Among these, sigmoid, tanh and ReLu are common.

4.2.1 Activation Functions

The purpose of an activation function is to produce an output from the weighted sum of all the inputs. We discuss a few of the most commonly used ones in this section. These have been proposed by researchers and practitioners over many years. Many other activation functions have been proposed although these are sufficient for most current users of artificial neural networks.

4.2.1.1 Sigmoid or Logistic Activation

The sigmoid or the logistic activation has been used in neural networks from the early years. It is not commonly used in modern neural networks. It is discussed here for historical reasons and to illustrate some desired properties of activation functions. The sigmoid function $\sigma(\ )$ is defined as

$$\sigma(x) = \frac{1}{1 + e^{-x}}, \tag{4.3}$$

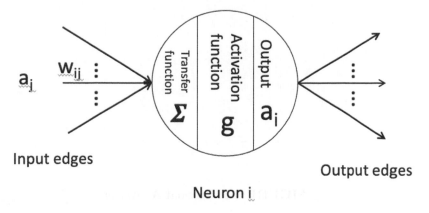

Input edges

Output edges

Neuron i

FIGURE 4.4: Structure of an Artificial Neuron

where x is its input. In the case of a neuron, it is the sum of all weighted inputs to the neuron.

As we see, at the origin, $x = 0$, and

$$\sigma(x) = \frac{1}{1 + e^{-0}} = \frac{1}{2} = 0.5.$$

At high values of x,

$$\lim_{x \to \infty} \sigma(x) = \lim_{x \to \infty} \frac{1}{1 + e^{-x}} \approx 1.$$

As seen in Figure 4.5, any value of x above 3 or 4, can be considered high. A value above 5 or 6 can be considered very high. At low values of x

$$\lim_{x \to -\infty} \sigma(x) = \lim_{x \to -\infty} \frac{1}{1 + e^{-x}} \approx 0.$$

Any value below -3 or -4 can be considered low.

The behavior of the sigmoid function is shown in Figure 4.5. The value of the output of the sigmoid activation function is zero or close to zero when the value of x or the sum of weighted inputs is large. The value of the output rises slowly and when $x = 0$, the output is exactly $\frac{1}{2}$, and then rises slowly to 1. Once it reaches a value close to 1, it is said to be saturated where the rise is real slow. In particular, when the value of x, the weighted sum of inputs is high or low, the rate of change of the function becomes very small. We will soon learn that ANNs learn by leveraging the rate of change or gradient of an activation function. When the gradient is low or almost zero, an ANN learns very slowly or does not learn at all. This issue is called the *vanishing gradient* problem. Thus, a neuron that uses sigmoid activation suffers from its inability to learn any further at saturation.

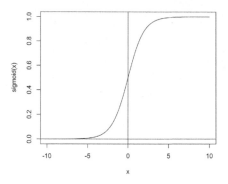

FIGURE 4.5: Sigmoid Activation

We also note that the sigmoid function is non-linear. This is an attractive property since it helps ANNs learn non-linear functions, which are more versatile than linear functions in mapping inputs to outputs. The non-linearity gets more pronounced when we place neurons in sequence. In other words, when an ANN learns a highly non-linear function, it is able to map inputs to outputs in very complex ways, which may be sometimes necessary to solve difficult problems in artificial intelligence.

4.2.1.2   *tanh* **Activation Function**

Another activation function that has been used frequently is *tanh*( ) given as

$$tanh(x) = \frac{e^x - e^{-x}}{e^x + e^{-x}}. \tag{4.4}$$

It is shown in Figure 4.6. Clearly, it is a non-linear function as well, allowing the learning of non-linear input to output mapping. When a number of *tanh* units are used in sequence, it leads to strong non-linearity.

The *tanh* function can also be written as

$$
\begin{aligned}
tanh(x) &= \frac{e^x - e^{-x}}{e^x + e^{-x}} \\
&= \frac{e^x - \frac{1}{e^{-x}}}{e^x + \frac{1}{e^{-x}}} \\
&= \frac{e^{2x} - 1}{e^{2x} + 1}.
\end{aligned} \tag{4.5}
$$

Some like this form of *tanh*( ) better because it does not use negative exponents. The, *tanh*( ) and *sigmoid*( ) functions are related. This relationship is

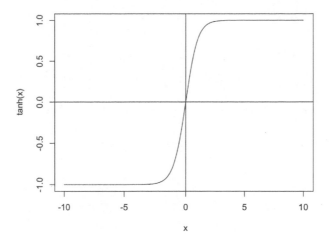

FIGURE 4.6: *tanh* Activation

shown clearly below.

$$tanh(x) = \frac{e^x - e^{-x}}{e^x + e^{-x}}$$

$$= \frac{1 - e^{-2x}}{1 + e^{-2x}}$$

$$= \frac{2 - 1 - e^{-2x}}{1 + e^{-2x}}$$

$$= \frac{2}{1 + e^{-2x}} - \frac{1 + e^{-2x}}{1 + e^{-2x}} \qquad (4.6)$$

In other words,

$$tanh(x) = 2\sigma(2x) - 1. \qquad (4.7)$$

Thus, the two functions, $\sigma(x)$ and $tanh(x)$ behave quite similarly. One can be considered a scaled and shifted version of the other. This means that any function that can be approximated by the sigmoid function can be approximated by the $tanh()$ function as well, and vice versa. We should note that the sigmoid function saturates at 0 and 1 at the two ends of its range, whereas the $tanh()$ function saturates at -1 and 1 at the two ends of its range. The value of $\sigma(x)$ is centered at 0.5 when $x = 0$, whereas $tanh(x) = 0$ at $x = 0$. Sometimes, $tanh(\ )$ is helpful because its value is centered at 0. The values of $tanh(x)$ are larger compared to $\sigma(x)$, and in particular, the rate of change or gradient is twice as strong. As a result, when learning with gradients, $tanh$ activation is likely to fare better. However, $tanh$ activation also suffers from the vanishing gradient problem. In modern neural networks, the $tanh(\ )$ activation function is used in long short-term memory (LSTM) networks.

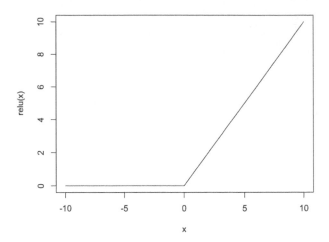

FIGURE 4.7: *relu* Activation

4.2.1.3   Rectified Linear Unit

The rectified linear unit is the most commonly used activation function for neural networks at this time. It is defined as

$$rect(x) = max(0, x) \qquad (4.8)$$

where x is the input to a neuron. It is also known as the ramp function. It is shown in Figure 4.7. Clearly, it is not a non-linear function in the traditional sense. It is a so-called piecewise non-linear function, composed of two simple linear functions, one on each side of the origin. On the negative side, it is simply a constant function $y = 0$. On the positive side of the origin along x, it is the simplest linear function possible, $y = x$. In spite of its simplicity, the ReLU function has been shown to be extremely effective as an activation function in neural networks, especially in so-called deep neural networks, where we can have up to hundreds of thousands of neurons, structured following a certain scheme or architecture, and millions or even billions of weighted connections or parameters. In fact, till about ten years ago, sigmoid and tanh were the workhorses of ANNs. This all changed with the introduction of the ReLU activation function. It was shown empirically that it trains much faster; one reason is that it is a simpler function compared to tanh and sigmoid, which requires computing powers of e.

As we have noted, the ReLU function's left hand side is a constant function with a value of 0 and a gradient of 0 as well. This makes it difficult to learn if a unit is stuck in this region. The problem is similar to the vanishing gradient problem faced by tanh and sigmoid functions. However, on the right side, the gradient is constant and is equal to 1; this gradient value is much larger than the gradients obtained from sigmoid or tanh. This facilitates faster learning, as learning depends on the magnitude of the gradient. However, a large value of the gradient, along with the unbounded value of the activation on the right hand side may cause the gradient to become large, making for very fast and eager learning, which is not so good either. In spite of such potential problems, ReLU is the dominant activation function in current practice of ANNs.

4.3 Simple Feed-Forward Neural Network Architectures

A single artificial neuron is interesting, but it cannot solve any interesting problem alone. Like real neurons, the computational power arises when a number of artificial neurons are put together to form a network. That is, an input is processed by not one single neuron, but a number of neurons in succession. The output of the last set of neurons is the output of the network. This allows for complex non-linear transformation of the input to the output. Neural networks actually learn these complex transformations—that is where their enormous power and usefulness come from.

When a network is built from neurons, the network is not put together arbitrarily like a regular graph or network that we may have seen. Rather, the structure is usually arranged in some simple ways, following the historical development of ANNs. There are hundreds of different architectures, but they all follow a few organizational patterns. Neurons are usually laid out in terms of layers. A layer is simply a number of neurons set up in what can be thought of as a sequence.

A layer can be thought of as sitting above another layer, unless it is the input layer. There is always an input layer, where the original input or signal values are received by the network. There is also an output layer where the network's outputs are extracted for interpretation and use. In between the input and output layers, there are layers that are called hidden layers that play roles in the complex non-linear transformation or mapping of the input to the output. In traditional neural networks, i.e., networks built till about ten years ago were shallow, with usually one or two hidden layers only. However, the situation has changed drastically in current neural networks, where there may be tens or even a hundred hidden layers. Due to increased depth, neural networks are also called deep neural networks. It is unclear what "deep" means; however, any network that has more than a few, say 3 or 4 hidden layers, can be called deep. We start our discussion of neural network architectures with very simple networks.

A simple neural network can have only two layers, an input layer and an output layer as shown in Figure 4.8. Assume the input layer has m nodes and the output layer has n nodes. The input layer has nodes that do not have any activation functions; they are simply activated to the value of the input

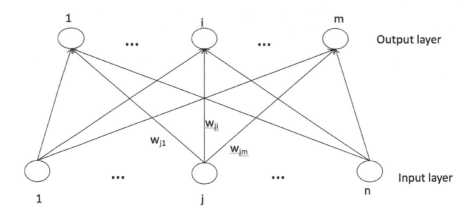

FIGURE 4.8: Two-layer Feedforward

component. In this case, the input is a vector length m. The output is a vector of length n.

Although it is possible to build a neural network with only an input and an output layer, no one really does so any more, although it used to be common several decades ago. These days, most neural network libraries require us to have at least one hidden layer.

Usually, all nodes in a layer have the same activation function and are thus, similar. These days, ReLU functions are used in the hidden layers. This is a feed-forwarrd neural network since the direction of flow of computation and information, when making a decision, is from input toward the output and is uni-directional.

Neural networks can be used to solve regression problems (discussed at length in Chapter 2) or classification (introduced in Chapter 3) problems. If we build a neural network for regression of one dependent variable, there may be only one output node. However, neural networks can be used to regress several dependent variables based on the same set of independent variables. If we have m output regressed variables, we will have m output nodes.

If an ANN is used for classification, the modern approach is to impose an additional layer on top of what would have been the output layer in the traditional sense. Suppose we are solving a classification problem with m output classes. In such a case, the output layer will have m output nodes. Based on our discussion so far, we know that the value output at any such node can potentially be any real value, positive or negative. Modern ANNs, almost always, normalize the output by superimposing another layer on top of the traditional output layer. The new output layer is called the softmax layer. Although softmax is called an activation function, it is not so in the normal sense in which we understand sigmoid, tanh or ReLU activation; these functions take the sum of weighted inputs coming to an individual neuron and

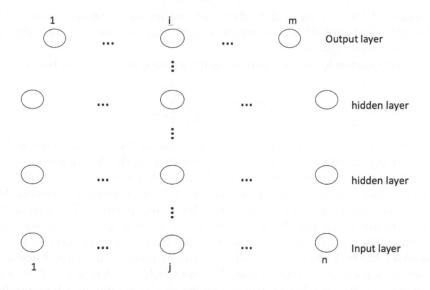

FIGURE 4.9: Regression with Neural Networks with k dependent variables and n input variables. Only 1 node at output if regressing one dependent variable.

produce an output for this individual neuron, without regards to other neurons in the layer it belongs to. In contrast, the softmax layer performs its computation by looking at the activations of all nodes in its layer as well as the layer from which it receives its input. The number of layers in the softmax layer is the same as the number of nodes in the layer from which it receives its input. The softmax layer normalizes its input in the sense that it ensures that the output of a corresponding node is scaled with respect to the values of the other nodes. The softmax layer achieves three requirements:

- The output of each softmax node is forced to be between 0 and 1. In other words, any negative values in input also become positive, although likely to be small.

- The output of all the nodes adds up to 1.

- Comparatively large values in the input layer have correspondingly large values in the softmax output, relatively speaking. However, the largest values become much larger in a relative sense, and the smallest values become much smaller in a relative sense. This allows clearer distinction between the winner(s) and the loser(s) in the softmax output.

Because of the first two points above, all outputs of a softmax layer can be thought of as probabilities to which a particular example belongs, after an ANN has been trained.

The equation for the softmax normalizing activation is given below:

$$softmax^{(i)} = \frac{e^{x^{(i)}}}{\sum_{j=1}^{k} e^{x^{(j)}}}. \tag{4.9}$$

Equation 4.9 relates the value of the ith input to the softmax layer and its corresponding output. Because of the summation in the denominator, we can clearly see normalization taking place. In both the numerator and the denominator, values are used to exponentiate $e \approx 2.78$. Therefore, corresponding values become large if the activation is large and positive. The sum in the denominator is constant for all values of i. A large input value becomes predominant in the numerator, and as a result, leads to a high softmax output value for the node. Since $e > 1$, raising e to any number, positive or negative, leads to a positive softmax output. We can easily see that if $x > 1$, e^x rises quickly as x becomes bigger, and e^x can become very large as x becomes big.

For negative values of x, e^x remains between 0 and 1, not changing fast, and rising very slowly from $x = -\infty$ to $x = 0$. Without going into any calculus or algebra, it is clear that the softmax transformation makes bigger activations stand out.

4.4 Training and Testing a Neural Network

Without going first into details of how an ANN learns, we will present a few simple examples of programatically training and testing neural networks. This will help the student make the idea of ANNs concrete and show that they are useful and powerful learners from data, motivating a desire to understand how they actually learn.

4.4.1 Datasets Used in This Chapter

Machine learning is an empirical area and needs datasets to demonstrate how algorithms and approaches work. In the context of ANNs, the datasets are usually large and represent "raw" data. Raw here means that ANNs are not usually fed featured datasets—datasets for which "experts" have figured out what characteristics of data examples are important. In computer vision, this may be the presence of straight lines, edges and particular shapes in particular locations within an image; these are usually obtained in a previous, possibly expensive, pre-processing step. For ANNs, the pre-processing step, if needed,

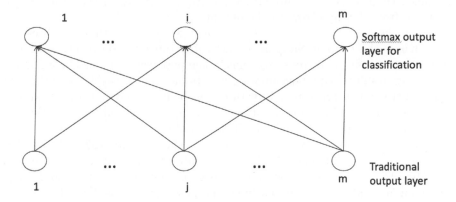

FIGURE 4.10: Softmax Layer in Classification Neural Networks

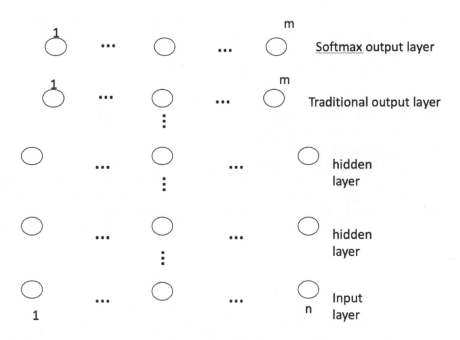

FIGURE 4.11: A modern Classification Neural Network. Note the presence of a softmax layer on top. The number of nodes in the penultimate (traditional output) layer and the softmax layer are the same.

is much simpler. ANNs, automatically discover features of the input as the data go from the input toward the output.

Like the previous chapters, we present a few datasets that are discussed throughout this chapter. Since neural networks, especially deep networks, have been used to solve numerous problems in the past few years, producing state-of-the-art results, many benchmark datasets have been produced so that the capabilities of novel neural architectures can be measured and compared, and improvements can be made. A dataset that is used to compare solutions to problems in computer vision is the MNIST dataset.

We work with this well-known dataset called the *MNIST* dataset. The MNIST database (Modified National Institute of Standards and Technology database) is a large database of handwritten digits that is commonly used for training and testing various image processing systems. The database is also widely used for training and testing in general machine learning. It was created by "re-mixing" the samples from NIST's original datasets. The creators felt that since NIST's original training dataset was taken from American Census Bureau employees, while the original testing dataset was taken from American high school students, they were not well-suited for machine learning experiments. The MNIST database contains 60,000 training images and 10,000 testing images. It is a subset of a larger set available from NIST.

The original black and white (bilevel) images from NIST were size normalized to fit in 20 × 20 pixel boxes while preserving their aspect ratio. The resulting images are given in terms of grey levels. An image was centered in a 28 × 28 image space by computing the center of mass of the pixels, and translating the image so as to position this point at the center of the 28 × 28 field. The 60,000 pattern training set contained examples from approximately 250 writers. The sets of writers of the training set and test set were disjoint. The test set has 10,000 patterns.

Figure 4.12 shows a few sample digits from the MNIST dataset. If we look at one of these samples, it is a two-dimensional 28 × 28 matrix, with 784 pixels in it. If we write out the matrix as a vector, with the 28 pixels of the second row following the 28 pixels of the first row, the 28 pixels of the third row following the 28 pixels of the second row, etc., we have a vector that is 784 long. So, the training and test datasets are 2-D matrices of size 60,000 × 784 and 10,000 × 784, respectively. The training and test labels are 1 × 10 vectors having a single 1 for one particular digit from 0 to 9, and 0 values for all other digits. A precise specification of the dataset is given in Table 4.1. The MNIST dataset can be downloaded from various websites. It is so commonly used in training and testing deep learning neural networks that many ANN libraries have functions that can download it from the Internet to be used within programs. In addition, CSV versions of the MNIST dataset are also available on the net since the original MNIST dataset is somewhat difficult to manipulate.

FIGURE 4.12: Sample MNIST Digits

TABLE 4.1: MNIST Dataset Details

train_x	60,000 × 784 uint8 (containing 60,000 training samples of 28 × 28 images each linearized into a 1 × 784 linear vector)
train_y	60,000 × 10 uint8 (containing 1 × 10 vectors having labels for the 60,000 training samples)
test_x	10,000 × 784 uint8 (containing 10,000 test samples of 28 × 28 images each linearized into a 1 × 784 linear vector)
test_y	10,000 × 10 uint8 (containing 1 × 10 vectors having labels for the 10,000 test samples)

4.4.2 Training and Testing a Neural Network in R Using Keras

There are several packages in R that allow us to build artificial neural networks. Of these, H20 and Keras are well-known. In this book, we discuss Keras, which is originally a Python package, but also available from within R. Since Keras is written in Python, we must also install Python if we do not have it already installed. The `reticulate` library enables interoperability between R and Python, allowing us to use Python libraries and Python objects. Lines 1-4 in the code, repeated below, from Section 4.4.2.5 are necessary to be able to work with Keras in R. Note that libraries keep on changing frequently, and hence variations to these commands may be necessary for proper working. For example, the current version may be different. In addition, file locations may be different.

```
#Make sure the libraries are installed. If not install them.
use_python("/usr/local/opt/python/bin/python3.6")
library (reticulate)
library (keras)
```

4.4.2.1 Loading the MNIST Dataset

When we install Keras, it automatically comes with routines that we can use to download some commonly used datasets from websites. One of these datasets is the MNIST dataset described above. We need to import this dataset into our program. Keras also provides a function that helps us load the data into the program. In our case, line 7 of the program given in Section 4.4.2.5 interfaces with Python to load the data into the program.

```
#get the MNIST dataset
mnist <- dataset_mnist()
```

Loading the dataset makes the data available in terms of two matrix components mnist.train and mnist.test. Each of these also have components in turn, x and y. Lines 10-14 separate the loaded MNIST dataset into its components as discussed.

```
#Set up training and testing subsets
x_train <- mnist$train$x
y_train <- mnist$train$y

x_test <- mnist$test$x
y_test <- mnist$test$y
```

In this case, x_train happens to be a uint8 (unsigned 8-bit integer) 3-D array (or matrix or tensor) with dimensions $60,000 \times 28 \times 28$; or in terms of Keras, it is a tensor with shape (60000, 28, 28). A tensor is simply another term for an array or matrix of various dimensions. x_test happens to be uint8 array image data with dimensions $60,000 \times 28 \times 28$ or a tensor of shape (60000, 28, 28). The y values are the class labels, which are in the range $0-9$. As a result y_train and y_test happen to be 1-dimensional uint8 array of digit labels. In Keras terms, y_train and y_test are tensors with shape (60000).

4.4.2.2 Reshaping the Input

Each layer in Keras has an input shape and an output shape. Keras automatically sets the input shape of a layer as equal to the output shape of the previous layer, but for the first layer, the programmer needs to give the input shape as a parameter. The MNIST dataset examples are in *width* $\times$ *height* pixels format. However, Keras expects each example to be a vector of values. That is why the input needs to be reshaped using a Keras function called reshape. The following two lines of code take each training and test example and make a 784 long vector out of each. The number of rows does not change.

```
x_train <- array_reshape(x_train, c(nrow(x_train), 784))
x_test <- array_reshape(x_test, c(nrow(x_test), 784))
```

The two statements above format the inputs as needed by Keras. However, when a network is trained, i) the network needs to be given the inputs at its input layer, which has 784 nodes and ii) at the same time, it needs to be given the output that is to be produced as well. The output layer has 10 nodes, and out of these, only one of the nodes needs to be 1 and the others need to be 0. In other words, at the output layer, we need a vector of length 10 where only one of the values is 1. This is called a *1-hot* vector.

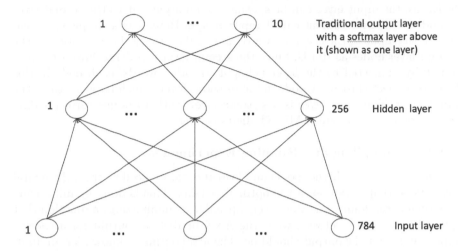

FIGURE 4.13: 3-Layer ANN for the MNIST dataset

```
y_train <- to_categorical(y_train, 10)
y_test <- to_categorical(y_test, 10)
```

The `to_categorically` function in Keras does this conversion of a single integer value representing the class of an example to a 1-hot vector. We need to do this for the y values for training examples as well as test examples.

4.4.2.3 Architecture and Learning

Let us first look at the kind of neural network architecture we want to build, and the kind of learning we want to happen in this network. We have discussed a simple feed-forward architecture in Section 4.3 already. The network we want to build has 784 input nodes, corresponding to each pixel in a data example. These are named i_1 through i_{784} in the network diagram in Figure 4.13. The intermediate or hidden layer has 512 nodes. The sub-network from the input layer to the hidden layer is *dense*, i.e., every node in the hidden layer is connected to every node in the input layer. The output layer has 10 nodes corresponding to the 10 classes we have to classify the digits into.

The program uses the variable `modelMNIST` to build the model, using the following statements.

```
#Set up the neural model
modelMNIST <- keras_model_sequential ()
modelMNIST <- layer_dense (modelMNIST, units=256, activation='relu',
                           input_shape=c(784))
modelMNIST <- layer_dense (modelMNIST, units=10, activation='softmax')
```

The first line says that we build a sequential or feed-forward network. The second line says that there is a dense or fully-connected layer that has 256 nodes.

In Keras, the input layer can be specified as an argument to the second layer of nodes with an argument called `input_shape`. Here, it is a simple specification of the input to the second layer of nodes. We specify the activation of the second layer nodes as ReLU. Since the second layer is a dense layer of nodes, it is fully-connected to the layer below it. In other words, each node in this layer is connected to every node in the previous layer, which happens to be the input layer. The third layer is a softmax layer, with 10 nodes corresponding to the ten classes to which MNIST digits belong.

4.4.2.4   Compiling and Running Experiments

Once we have built the model and loaded data, we need to train the neural network so that it can learn the optimal weights. This is done by initializing the weights randomly, and iteratively optimizing (minimizing) a function that quantifies the error between what the ANN predicts as output for a specific input, and what the output should be. The ideal output is known for an input example since we are working on supervised learning. Weights are updated at the end of a batch. The error function is usually called a loss function in ANNs, and we need to specify the loss function we want to use. We also need to specify the metric we use to evaluate the quality of learning. This we can do by compiling our model. In particular, to compile the model, we need to specify a loss or error function and an optimizer. We discuss loss functions and optimizer choices later in this chapter.

To learn, an ANN uses an algorithm called backpropagation. To learn using backpropagation, we can use a variety of error or loss functions as well as a variety of optimization algorithms. In particular, in our program, we use `mean_squared_error` as the loss or error function and `RMSprop()` as the optimizer. For each example during training, the mean squared error is computed between the results produced by the neural network and the actual expected outputs at output nodes. Backpropagation also obtains derivatives of the activation functions at the output nodes and sends the error that involves derivatives back along all edges, making corrections progressively backwards. We discuss backpropagation in detail in Section 4.5. Compilation configures our model in the sense that it puts everything together and makes it ready for training.

```
1  #Compile the model, i.e., set it up to run
2  compile(modelMNIST, loss = 'mean_squared_error', optimizer = optimizer_rmsprop(),
3      metrics = c('accuracy'))
```

As mentioned before, the MNIST dataset has 60,000 training examples. When we run a machine learning algorithm, an optimization algorithm is run, considering errors made over the entire dataset, in batches. The goal is to find an optimal set of weights for the edges based on the training data. The process in which ANN goes over all the examples in the training dataset once is called an *epoch*. In some machine learning algorithms, the learner runs over the data only once. It is not so in the case of ANNs. In general, the ANN goes over the dataset many times to train well. It is quite likely that the ANN has to run through the dataset tens of times to produce good results.

The ANN can go over one example at a time and update the hundreds or thousands or millions of weights it has. If this is the way the experiment is run, we say that the batch size is 1. If it goes over all the examples, say 60,000 in the case of the MNIST dataset, before updating weights, we say that the batch size is 60,000. However, as the reader may have surmised, running over one example at a time and updating all the weights for each example, and doing it 60,000 times in an epoch makes the learning process slow. On the other hand, a batch of size 60,000 requires the program to store results of a lot of intermediate computations, making memory requirement high. Thus, either one of these batch sizes is usually not appropriate. A batch size like 128, where the updates to weights take place after going over 128 data examples may be a better choice. When a smaller batch size, say 64 or 128 is used, a batch is usually called a mini-batch. We use the term batch for mini-batch as well.

Since updates are performed after processing a batch of data, the effects of the individual examples need to be stored till the end of the batch. This requires memory. The individual weight updates computed for an edge are averaged for a batch to actually update the weight of the edge. The weight of each edge is updated after a batch. If the batch size is large, a lot of memory may be needed. However, if the batch size is small, the amount of time required may be high. So, this is a balancing act.

The number of epochs and batch size can be set in Keras by assigning values to the appropriate variables in the `fit` command. One should change these parameter values and perform experiments to see if the results and time taken vary substantially.

Once the model has been compiled, it can be trained. To train, we have to give the training data, the testing data, the batch size and the number of epochs to train. We can also specify how the data should be split for training and validation. In this case, 80% of the data is used for training and 20% for validating. Note that validation subset of data is different from the testing subset. The model should not have been trained on the validation data.

```
1  #the fit function runs the model. We save the run history in a variable
2  history <- fit(modelMNIST, x_train, y_train,epochs = 10, batch_size = 128,
3       validation_split = 0.2)
```

4.4.2.5 The Entire Program

In the previous subsections, we discussed the various components that make a simple program using Keras in R. The parts are once again listed below:

- Install and import libraries necessary.
- Load the dataset.
- Split the dataset into training and testing subsets.

- Reshape the dataset as necessary.

- Build the ANN model.

- Set up the experimental protocol (procedure).

- Run the experiments.

- Evaluate results of experiments and print results.

All these parts are put together into a working program, which is shown below. It creates a 3-level ANN with backpropagation to classify the MNIST dataset into ten classes.

```
1  #Make sure the libraries are installed. If not install them.
2  use_python("/usr/local/opt/python/bin/python3.6")
3  library (reticulate)
4  library (keras)
5
6  #get the MNIST dataset
7  mnist <- dataset_mnist()
8
9  #Set up training and testing subsets
10 x_train <- mnist$train$x
11 y_train <- mnist$train$y
12
13 x_test <- mnist$test$x
14 y_test <- mnist$test$y
15
16 x_train <- array_reshape(x_train, c(nrow(x_train), 784))
17 x_test <- array_reshape(x_test, c(nrow(x_test), 784))
18 x_train <- x_train / 255
19 x_test <- x_test / 255
20 y_train <- to_categorical(y_train, 10)
21 y_test <- to_categorical(y_test, 10)
22
23 #Set up the neural model
24 modelMNIST <- keras_model_sequential ()
25 modelMNIST <- layer_dense (modelMNIST, units=256, activation='relu', input_shape=c(784))
26 modelMNIST <- layer_dense (modelMNIST, units=10, activation='softmax')
27
28 #print a summary of the model
29 summary(modelMNIST)
30
31 #Compile the model, i.e., set it up to run
32 compile(modelMNIST, loss = 'categorical_crossentropy',
33     optimizer = optimizer_rmsprop(),metrics = c('accuracy'))
34
35 #the fit function runs the model. We save the run history in a variable
36 history <- fit(modelMNIST, x_train, y_train, epochs = 10, batch_size = 128,
37     validation_split = 0.2)
38
39 #plot how training and testing went
40 plot (history)
41
42 #evaluate the model's performance on test data
43 evaluate(modelMNIST, x_test, y_test)
```

As we run this program, it trains the ANN model on the training data and tests it on the test data. It produces a lot of output on the screen, which we do not show here. Below, we show the output of the summary(modelMNIST) statement.

```
Model: "sequential_8"

Layer (type)                   Output Shape               Param #
=================================================================
dense_13 (Dense)               (None, 256)                200960

dense_14 (Dense)               (None, 10)                 2570
=================================================================
Total params: 203,530
Trainable params: 203,530
Non-trainable params: 0
```

The model summary above shows that there are $203,530$ parameters or weights to learn in the simple model we have built. It goes through 30 epochs of learning, and it prints out the accuracy values obtained in each run. It finally provides an average of the accuracy values obtained over the 10 runs. The values of the loss function and accuracy as the ANN goes through epochs are shown in Figure 4.14

4.4.2.6 Observations about Empirical ANN Learning

Changing the Number of Epochs: There are two graphs in Figure 4.14—the top one is a loss graph, and the second is an accuracy graph. The loss graph shows how the cumulative error in classification changes as the number of epochs rises. As mentioned earlier, a showing of all examples in the training set once to the ANN classifier along with the correct labels and making updates to

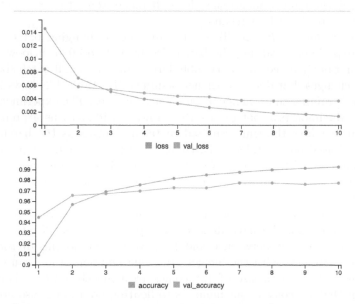

FIGURE 4.14: Results from a 3 layer feed-forward ANN for classification of MNIST Digits, trained for 10 epochs. Upper: Loss over epochs; Lower: Accuracy over epochs.

the weights is called an epoch. The weights of the ANN are usually initialized randomly, and therefore, the network is likely to have large cumulative errors in early epochs. The ANN learns by making adjustments to the weights with a view to lowering the error or loss, using backpropagation. That is why we see that the loss comes down as we train the ANN on an increasing number of epochs. Initially, the reduction in loss is quite steep and then it slows down. There are two losses shown: training loss and validation loss. The two losses are computed once in each epoch. The graph shows that training loss continues to go down, albeit slowly, till epoch number 10. This means that as we train the ANN more and more, it adjusts the weights to perform better and better on the training set. However, on the validation subset of data, the loss comes down till epoch 7 and then remains more or less constant. This means that the network overfits after epoch 7, possibly even after epoch 5. Overfitting means that the network does well on the training data, but does not do so well on data on which it was not trained. So, the training could have possibly stopped after epoch 7. Note that the network has not been tested yet on the actual test data, which we had kept separately in the beginning of the program.

The lower graph in Figure 4.14 gives us how the measure of goodness, which is accuracy here, changes as the number of epochs increases. Two accuracies are plotted here, training accuracy and validation accuracy. Accuracy on the training data keeps on increasing till epoch 10 although the rate of growth slows down after the first 4 or 5 epochs. The validation epoch keeps on rising unevenly after the first two epochs, and the rise is slow thereafter. In this case, even though the improvements slow down, it makes sense to continue training till the 10th epoch for best results.

Changing the Batch Size: To see the impact of changing batch size, we perform experiments varying the batch size from 1 to 4096, in powers of 2. The amount of time becomes noticeably larger for very small batch sizes. The accuracy changes, but does not change significantly. Table 4.2 shows how accuracy and training time change as batch size changes. The experiments were run on a Macbook pro, not on a GPU. However, it gives us a pretty good idea regarding how the time required to train changes as the batch size is increased. Based on the table, a batch size of 128 is a good compromise between accuracy and time taken. Note that as machines become more powerful over the years, the actual times required will become lower, but the trend is unlikely to change.

Normalization of Data: Each pixel is in grayscale and is a single value between 0 and 255. We have divided each pixel value by 255 to obtain a floating point number between 0 and 1. If we do not do this division, and perform experiments with the integer grayscale numbers, the accuracy after 10 epochs with a batch size of 128 comes to a measly 0.2125. Thus, it is important that we convert our numbers to floating point, and also normalize the numbers to a small range of floating point numbers, before using them with ANNs.

TABLE 4.2: Time Taken and Accuracy as the Batch Size rises

Batch Size	Accuracy	Time/sample in nanosec
1	0.9544	3000
2	0.9700	1000
4	0.9729	734
8	0.9732	347
16	0.9759	183
32	0.9763	93
64	0.9737	56
128	0.9783	38
256	0.9725	27
512	0.9699	19
1024	0.9678	17
2048	0.9493	16
4096	0.9450	14

Using Different Activation Functions and More Layers: In the first ANN we have discussed so far, we use a ReLU layer, followed by a softmax layer. This architecture follows the current trend in neural networks, where the hidden layers usually use ReLU activation. The hidden layers are followed by usually a so-called multi-layered perceptron, which has one or more layers that use sigmoid activation. The last sigmoid activation layer is followed by a softmax layer, that actually performs the classification. In the example discussed earlier, we have one ReLU layer followed by a softmax layer; so, it is one of the simplest classification architectures we can build. Table 4.3 shows several architectures and the corresponding accuracies. In this problem, changes to architectures or activation functions do not make a whole lot of difference, the results being consistently high. The current state-of-the-art in MNIST classification can be found in its leaderboard, and is almost 100%[1].

Using Different Loss Functions: In machine learning, whether its ANN or other algorithms, the "modern" way to train is by optimizing a loss or error function across all labeled examples.

As we train a neural network, we can compute the error at an output node in various ways. The term *loss* is commonly used for errors. A simple way to compute the error or loss at an output node for a training example $\vec{x}_i$ may be to simply compute the difference between t_i, the expected or target value for this example, and y_i, the actual output produced by the neural network with the weights it has at this time. So, such an error at the output node can be written as $t_i - y_i$. We can obtain a cumulative or total raw error over all the examples in the training set by adding up all the errors. So, the total error or

---

[1]https://paperswithcode.com/sota/image-classification-on-mnist

TABLE 4.3:   Changing Activation Functions and Layers with Epochs 10, Mean Squared Error and RMSProp, Normalized Data

Architecture	Accuracy
input(784), relu(256), relu(128), sigmoid(16), softmax(10)	0.9773
input(784), relu(256), relu(128), relu(16), softmax(10)	0.9739
input(784), relu(512), relu(256), relu(128), relu(64), relu(32), softmax(10)	0.9737
input(784), relu(512), relu(256), relu(128), sigmoid(16), relu(32), softmax(10)	0.9779

loss at this output node for the entire dataset can be written as

$$\mathcal{L}_{sumraw} = \sum_{i=1}^{N}(t_i - y_i), \tag{4.10}$$

and the mean raw error can be obtained as

$$\mathcal{L}_{meanraw} = \frac{1}{N}\sum_{i=1}^{N}(t_i - y_i), \tag{4.11}$$

where N is the total number of examples in the training set. However, the error at an output node can be positive or negative, and if we compute the total (or mean) of such errors over all the training examples, some positive errors may cancel out negative errors, making the total (or mean) smaller that it really is. Therefore, instead of computing the sum (or average) of raw errors, we can compute the sum of squared raw errors or the mean of squared errors as the loss or error that we want to reduce by changing the weights of the network.

$$\mathcal{L}_{sumsquared} = \sum_{i=1}^{N}(t_i - y_i)^2, \tag{4.12}$$

and the mean squared error can be obtained as

$$\mathcal{L}_{meansquared} = \frac{1}{N}\sum_{i=1}^{N}(t_i - y_i)^2. \tag{4.13}$$

The last one, given in Equation 4.13 is often used in neural network training. It is also available in Keras.

The well-known backpropagation algorithm attempts to reduce the mean squared error between observed outputs and expected outputs (or labels given to the examples) during training. Among the loss functions Keras provides, Mean Squared Error and Multiclass Logarithmic (also called Categorical Cross Entropy) are commonly used. Categorical Cross Entropy Loss is usually used when we have multiple classes like in our model where we have ten different

TABLE 4.4: Changing Loss Functions in a 3-layer Network: 784 Input, 256 ReLU, and 10 Dense with Softmax (Output) with Batch Size 128, Epochs 10, RMSProp

Loss Function	Accuracy
Mean Squared Error	0.9565
Categorical Cross Entropy	0.9600

TABLE 4.5: Changing Optimizers in a 3-layer Network: 784 Input, 256 Dense with ReLU, and 10 Dense with Softmax (Output) with Batch Size 128, Epochs 10, Mean Squared Loss

Optimizer	Accuracy
RMSProp	0.9565
Stochastic Gradient Descent	0.1387
Adam	0.9582

classes for the ten digits. We do not discuss details of these loss functions in this book.

We run some experiments with our simple 3-layered network with these three loss functions, with 784 input, 512 dense (Layer 2), and 10 dense (Output) nodes with Batch Size 128, Epochs 20, with RMSProp as the optimizer. Table 4.4 shows the results. We perform normalization as discussed earlier.

Using Different Optimizers: Modern approaches to ANNs can use different types of optimization functions to find the weights that optimally fit the training data. Among various possible optimizers are Root Mean Square Optimizer, Stochastic Gradient Descent and Adam, although there are several others available. We run experiments with three different optimizers, otherwise keeping the parameters the same as before. We do not discuss various optimizers in this book except (Stochastic) Gradient Descent. We run our experiments with default parameter values. In the results reported throughout this document, we perform an experiment with a single set of parameters only once. Normally, the results come out a bit different each time, and hence, it is advised to run several times with the same set of parameter values and report an average result. Table 4.5 shows how the accuracy changes when we change the optimization functions for our MNIST example. The input values were normalized as discussed earlier. The validation accuracy is very low when SGD is used in combination with mean squared error in this case. However, with SGD, if we change the loss to categorical cross-entropy, the accuracy rises to 0.8258.

Reducing Overfitting: Overfitting occurs when the model fits the training data with a very small error, but cannot generalize well. In other words, it does not work so well with unseen test data. Because a trained classifier can be published and then people can use it on any appropriate but unseen data, the results may come out bad if it is not able to discount accidental regularities it may find in the training data, whether due to presence of noisy data or it so happens that some regularities not found generally occur in the training data. Several approaches can be used to address overfitting. One is the use of regularizers and the other is the use of Dropout in certain layers. We do not go deep into these topics in this book.

4.5   Backpropagation

4.5.1   Backpropagation at a Conceptual Level

In the previous sections, we have introduced ANNs and also have seen how we can use them to solve classification problems. For the MNIST dataset, which is frequently used in machine learning research, even simple neural networks give accuracy in the 97-98% range. More complex networks give up to 99.90% accuracy, as of early 2022.

Although we have seen how to program ANNs using Keras, we have not yet discussed how neural networks actually learn to perform well in classification and other tasks. This is the topic of this section.

A neural network is a layered layout of neurons. A layer below is connected to the layer immediately above in simple ANNs. There is an input layer at the bottom and an output layer at the top. Each connection from a node in a lower to a higher layer has a weight. A weight is also called a parameter. Even a simple network can have a few hundred thousand or a few million parameters. These parameter values need to be learned from the training dataset in an empirical manner so that the ANN can classify unseen examples. Initially, before the ANN is trained, the weights are usually set randomly to some small values. As training examples are shown to the network, it learns to adjust the weights a bit so that it possibly does a little better at classification the next time.

Let us assume for the time being, that training examples are shown one at a time. We also assume that all activations in all layers are sigmoid. The first example is shown to a neural network whose weights are random to start with. This network is asked to classify an example, for which the ground truth is also known. It is quite likely that this ANN, which has random knowledge, is going to make errors in classification. Note that even though the network is performing classification, it is generating a number at the output. For example, if the example being shown should have an output of, say 1, saying it belongs to a certain class, the ANN may produce another value, say 0.2. Thus, there is likely to be a numeric difference between what the ANN should produce as output, and what it actually produces. This can be used to compute an error or loss for the specific example. This individual example's loss is computed at the output node of the network. The loss at the output node can be used to compute corrections to the weights on the edges coming to the output node. In other words, the loss at the output node or layer can be backpropagated to the layer directly beneath it. It is now possible to perform computations so that the errors committed at the nodes in the penultimate layer (due to the error at the output, of course) are backpropagated further to the layer just before the penultimate layer. In a similar way, layer by layer, the error at the output can be backpropagated all the way to the layer just above the input layer. At each layer, based on the error at a certain layer, the weights on the edges coming up to the layer can be changed a little bit so that the next time around, the ANN may make a slightly smaller error.

Thus, the training of an ANN with respect to a training example happens in a bidirectional manner. First, the ANN considers the input and performs computations, using the current weights on the edges, in a forward manner layer by layer, all the way to the output. At the output, an error or loss value is computed, and the loss is backpropagated, layer by layer, in a backward manner, making corrections to all the weights in the network. Making corrections to all the weights for an example constitutes learning in an ANN. Such a network may have hundreds of thousands or millions of weights, and thousands or tens of thousands of neurons, each with its own activation function. It is quite computation-intensive, even to learn from one training example. The neural network learns from each example in the training set, and there may be tens of thousands or more such examples. Learning from all the examples once is called an epoch of learning, as we already know. Thus, an epoch of learning in an ANN is a very computation-intensive task.

In ANNs, one epoch of learning is not enough. Often, the ANN system needs to go through tens of epochs to learn properly. Usually, we can run it for a certain number of epochs till the cumulative loss for the epoch does not decrease any more. Often, the cumulative loss is dramatic for a few of the first epochs and then slows down, and finally levels off, or even becomes worse. The process of backpropagation is illustrated in Figure 4.15.

We should note that at the output layer, we may have not just one node, but several. For example, in a network that classifiers the MNIST dataset, there are ten output nodes. For a single training example, each output node computes a loss. The individual losses at the output nodes are used to make corrections to edge weights connecting back to the output layer. Usually there is a softmax layer at the end of a neural network, and hence different computations are necessary for backpropagation at the output layer.

4.5.2 Mathematics of Backpropagation

ANNs usually use a technique called Gradient Descent to learn weights from examples. Let an input training example be $\langle \vec{x}, y \rangle$, where $\vec{x}$ is a vector and y is the actual label on the example; the label is assumed to be number.

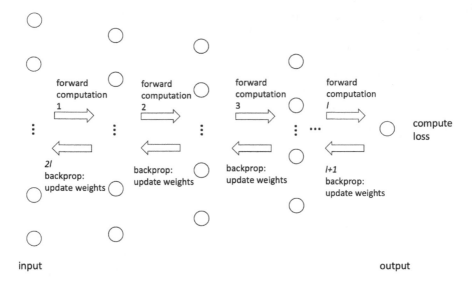

FIGURE 4.15: An ANN showing feedforward, error and weight update computations. These are followed by error computation at the output. Steps l+1 through 2l are backpropagation of error computations.

The number represents a "degree of probability of the example belonging to a specific class. For example, if the class corresponds to the digit 4, a value of 1 implies that the corresponding example definitely is a 4. If the value is 0, the example definitely is not a 4. For a value between 0 and 1, the example may be 4 with a degree between 0 and 1. In the case of the MNIST dataset, there are 10 output nodes corresponding to digits 0–9. Each will have a degree of belongingness to the corresponding class. The output node with the highest value determines the class to which the example belongs.

Let the output or the value predicted by the ANN at an output node for this specific training example be $\hat{y}$. Let the error or loss be squared loss, which we write as $\frac{1}{2}(y - \hat{y})^2$. The $\frac{1}{2}$ in front is simply to make the formulas come out a little simpler as we perform derivation.

At any non-input layer in the neural network, there are inputs, weights on the edges out of the inputs, a summation function—also called a transfer function sometime, an activation function, and an output produced by the activation function. The various components and steps in the processing that takes place at a generic node have been separated and clearly shown in Figure 4.16. In this diagram, the following are clearly identified.

- The inputs $x_1 \cdots x_n$ and the corresponding weights $w_{1j} \cdots w_{nj}$. The inputs to a neuron are the outputs of the previous layer of neurons. If the neuron is on the first layer after the input layer, the o_ks of the input layer are simply the inputs x_k to the network.

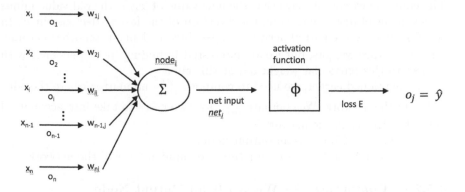

FIGURE 4.16: The steps in computing the output at node j

- Σ, the summation or the transfer function.
- net_j, the weighted sum of the inputs produced by the transfer function.
- ϕ, the activation function. We assume that the activation function is the sigmoid function $\sigma(z) = \frac{1}{1+e^{-z}}$ for the discussions that follow.
- o_j, the output of the activation function:

$$o_j = \phi(net_j) = \phi\left(\sum_{i=1}^{n} w_{ij} o_i\right).$$

- the loss or error function $E = \frac{1}{2}(y - \hat{y})$.

We focus on changing a generic weight w_{ij} based on E at the output node. Note that the edge ij can be a anywhere in the network, not just an edge incident on an output node. No matter where the edge ij is, after training on an example and obtaining error E, we want to compute the change $\triangle w_{ij}$ in the weight w_{ij} so that the new weight becomes

$$w_{ij} \leftarrow w_{ij} + \triangle w_{ij}. \tag{4.14}$$

$\triangle w_{ij}$ should be such that it reduces the error E the fastest. This can be accomplished by first computing the gradient of E with respect to w_{ij}. The gradient $\frac{\partial E}{\partial w_{ij}}$ gives us the direction in which E increases the fastest. Since our goal is not to increase E, but to decrease it the fastest, we change E in the direction of the negative of the gradient, given as $-\frac{\partial E}{\partial w_{ij}}$. How much we move in this direction depends on a parameter called the learning rate α. Therefore, $\triangle w_{ij} = -\alpha \frac{\partial E}{\partial w_{ij}}$. As a result, we can write Equation (4.14) as

$$w_{ij} \leftarrow w_{ij} - \alpha \frac{\partial E}{\partial w_{ij}}. \tag{4.15}$$

The equation essentially says that the new value of w_{ij} is the old value minus a movement of distance α along the direction of the fastest decrease in E. In this formula, we use partial derivative $\frac{\partial E}{\partial w_{ij}}$ since E depends on other parameters, e.g., other weights, but we are interested in finding the dependence with respect to this particular weight w_{ij} at this time.

As mentioned earlier, we need to compute $\frac{\partial E}{\partial w_{ij}}$ no matter where the edge ij lies in the network. We compute $\frac{\partial E}{\partial w_{ij}}$ for two cases to take into account all possible edges ij in the network.

> Case 1: Node j is an output node,
> Case 2: Node j is on an internal or hidden layer of the network.

4.5.2.1 Computing $\frac{\partial E}{\partial w_{ij}}$ When j Is an Output Node

We refer to Figure 4.16 to perform the computation, but we look at the figure backwards from error E to the input. We also use the chain rule in derivatives. The chain rule is used when there is a sequence of variables on which the variable of interest depends. In this case, we can say that E depends on o_j, which in turn depends on net_j, which in turn depends on w_{ij}. As a result, we can write

$$\frac{\partial E}{\partial w_{ij}} = \frac{\partial E}{\partial o_j} \frac{\partial o_j}{\partial w_{ij}}$$

$$= \frac{\partial E}{\partial o_j} \frac{\partial o_j}{\partial net_j} \frac{\partial net_j}{\partial w_{ij}} \tag{4.16}$$

Here, $\frac{\partial E}{\partial w_{ij}}$ is the rate of change (gradient) of E w.r.t. w_{ij}, and can be obtained by multiplying three rates of change: $\frac{\partial E}{\partial o_j}$, which is the rate of change of E w.r.t. o_j—this is the derivative of the loss or error function w.r.t. the output; $\frac{\partial o_j}{\partial net_j}$, which is the rate of change of o_j w.r.t. net_j—this is the derivative of the output of the activation function with the total input to the activation function; and, $\frac{\partial net_j}{\partial w_{ij}}$, which is the rate of change of net_j w.r.t. w_{ij}—this is the gradient of the sum of the weighted input to the transfer function w.r.t. one particular weight on an incoming edge to the transfer function. We have broken up the original gradient into these three gradients to show direct dependences, as indicated clearly in Figure 4.17.

Let us compute the three partial derivatives in Equation 4.16 one by one. We first compute the derivative of the loss function, $\frac{\partial E}{\partial o_j}$. If node j is at the

$$w_{ij} \longrightarrow net_j \longrightarrow o_j \longrightarrow E$$

FIGURE 4.17: Direct chain of gradient dependencies at node j in the calculation of E

output layer, $o_j = y$, the predicted output at node j. So, we can write

$$\frac{\partial E}{\partial o_j} = \frac{\partial}{\partial o_j}\left[\frac{1}{2}(y - \hat{y})^2\right]$$

$$= \frac{1}{2}\frac{\partial}{\partial \hat{y}}(y - \hat{y})^2$$

$$= \frac{1}{2} \times 2(y - \hat{y})$$

$$= -(y - \hat{y}) \tag{4.17}$$

Obviously, the value above depends on the error or loss function used. The loss function should be easily differentiable since this computation needs to be done for each training example for each epoch.

Let us now compute the second partial derivate,

$$\frac{\partial o_j}{\partial net_j} = \frac{\partial}{\partial net_j}\sigma(net_j).$$

This calls for finding the derivative of the activation function, which happens to be sigmoid in this case. Instead of writing $\sigma(net_j)$, let us write it as $\sigma(z)$ and obtain its derivative w.r.t. z.

$$\frac{d}{dz}\sigma(z) = \frac{d}{dz}\left(\frac{1}{1 + e^{-z}}\right)$$

$$= \frac{d}{dz}\left(1 + e^{-z}\right)^{-1}$$

$$= -1\left(1 + e^{-z}\right)^{-2}\frac{d}{dz}\left(1 + e^{-z}\right)$$

$$= \frac{e^{-z}}{(1 + e^{-z})^2}$$

$$= \frac{1}{1 + e^{-z}}\frac{e^{-z}}{(1 + e^{-z})}$$

$$= \frac{1}{1 + e^{-z}}\left(1 - \frac{1}{1 + e^{-z}}\right)$$

$$= \sigma(z)(1 - \sigma(z)) \tag{4.18}$$

Therefore, we can write

$$\frac{\partial o_j}{\partial net_j} = \sigma(net_j)(1 - \sigma(net_j))$$

$$= o_j(1 - o_j)$$

$$= \hat{y}(1 - \hat{y}). \tag{4.19}$$

Obviously, the value above depends on the activation function used. The activation function should be efficiently differentiable, since this computation also needs to be done for each training example for each epoch.

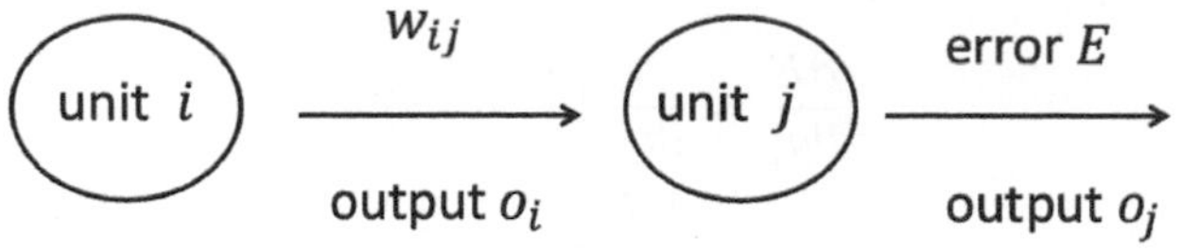

FIGURE 4.18: Simple view of the Backpropagation formula

Let us now compute the third partial derivative, $\frac{\partial net_j}{\partial w_{ij}}$.

$$
\begin{aligned}
\frac{\partial net_j}{\partial w_{ij}} &= \frac{\partial}{\partial w_{ij}}\left(\sum_{k=1}^{n} w_{kj}o_j\right) \\
&= \frac{\partial}{\partial w_{ij}}\left(w_{1j}o_1 + w_{2j}o_2 + \cdots + w_{ij}o_i + \cdots + w_{nj}o_n\right) \\
&= 0 + 0 + \cdots o_i + \cdots 0 \\
&= o_i
\end{aligned}
\tag{4.20}
$$

This computation is the easiest of the three gradient computations.

Having obtained all three partial derivatives in Equation 4.16, we can write

$$
\begin{aligned}
\frac{\partial E}{\partial w_{ij}} &= \frac{\partial E}{\partial o_j}\frac{\partial o_j}{\partial net_j}\frac{\partial net_j}{\partial w_{ij}} \\
&= (y - \hat{y})\hat{y}(1 - \hat{y})o_i \\
&= \delta_j o_i
\end{aligned}
\tag{4.21}
$$

where $\delta_j = -(y - \hat{y})\hat{y}(1 - \hat{y}) = (y - o_j)o_j(1 - o_j)$. This is specifically for the case when the activation function is sigmoid and the loss function is squared error. Thus, we can rewrite the weight update equation, Equation 4.15 as

$$
w_{ij} \leftarrow w_{ij} - \alpha\delta_j o_i
\tag{4.22}
$$

where α is the learning rate. δ_j depends on the actual output of the jth unit and the ideal output which should have been obtained at the jth unit, whereas o_i is the ith input to unit j.

In general, as seen in Figure 4.18, the change in weight w_{ij} depends on the derivative of the error function at unit j, and the activation at the input unit i. In other words, to perform backpropagation update to an edge's weight, we need to know the gradient of the error function at the ending node for the edge, and the value of the activation at the starting node. For the specific case when we have sigmoid activation, squared error loss, the update formula for a weight w_{ij} is given in Equation 4.22. If the activation function or loss function changes, we will have a different formula for weight update using backpropagation.

4.5.2.2 Computing $\frac{\partial E}{\partial w_{ij}}$ When j Is a Hidden Layer Node

Let the node j be situated in a layer just below a layer numbered L. A generic node in this layer be called l. In forward computation, the signal moves forward from j toward L, and then toward the output node, where we compute error or loss E.

In backpropagation, the error is backpropagated from the output layer to layer L, and then from L down through node j to the layer below, and then all the way to the back to the input layer. Consider the nodes in layer L. At each node in layer L, there is some error. These errors all contribute to the total error seen at node j. In fact, the error seen at node j is the sum of all the errors at each node in layer L.

$$E_j = \sum_{l \in L} E_l \tag{4.23}$$

Let us compute the first term in Equation 4.16. We get

$$\frac{\partial E}{\partial o_j} = \frac{\partial}{\partial o_j} \left(\sum_{l \in L} E_l \right)$$

$$= \sum_{l \in L} \frac{\partial}{\partial o_j} E_l$$

$$= \sum_{l \in L} \left(\frac{\partial E_l}{\partial o_l} \frac{\partial o_l}{\partial net_l} \frac{\partial net_l}{\partial o_j} \right)$$

$$= \sum_{l \in L} \delta_l w_{jl} \tag{4.24}$$

where δ_l is the product of the first two terms inside the parentheses, and $net_l = \sum_{j=1} w_{jl} o_j$.

Replacing the value of $\frac{\partial E}{\partial o_j}$ in Equation 4.16, we get

$$\frac{\partial E}{\partial w_{ij}} = \frac{\partial E}{\partial o_j} \frac{\partial o_j}{\partial net_j} \frac{\partial net_j}{\partial w_{ij}}$$

$$= \left(\sum_{l \in L} \delta_l w_{jl} \right) \frac{\partial o_j}{\partial net_j} \frac{\partial net_j}{\partial w_{ij}}$$

$$= \left(\sum_{l \in L} \delta_l w_{jl} \right) o_j(1 - o_j) o_i$$

So, finally the weight update equation is the one given below.

$$w_{ij} \leftarrow w_{ij} + \Delta w_{ij}, \tag{4.25}$$

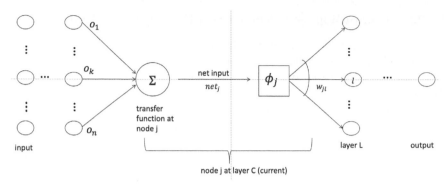

FIGURE 4.19: The steps in computing the activation at a hidden node j

where

$$\triangle w_{ij} = \alpha o_i \delta_j, \quad \text{and}$$

$$\delta_j = \begin{cases} (y - \hat{y})\hat{y}(1 - \hat{y}) \text{ if } j \text{ is an output node} \\ \left(\sum_{l \in L} \delta_l w_{jl} \right) o_j(1 - o_j) \text{ if } j \text{ is a hidden layer node} \end{cases}$$

Once again, the formula above is for the case when the sigmoid activation is used with squared error.

4.5.2.3 Backpropagation with Different Activation Functions

In the discussion in the previous section, we assumed that the activation function is sigmoid in all layers except of course, for the input layer. We also made the assumptiion that there is one output node, although we call it node j. We will still assume there is one output node, but assume that different layers may have different activation functions. Assume that we have an ANN where the hidden layers have ReLU activation, followed by a layer where we have sigmoid activation, followed by softmax activation, like the one shown in Figure 4.20. To be able to backpropagate, we need to be able to obtain derivations of the ReLU and softmax activations.

Backpropagation with ReLU The ReLU function is

$$r(x) = \begin{cases} x \text{ if } x > 0 \\ 0 \text{ otherwise.} \end{cases} \tag{4.26}$$

So, the derivative of the ReLU function is

$$r'(x) = \begin{cases} 1 \text{ if } x > 0 \\ 0 \text{ otherwise.} \end{cases} \tag{4.27}$$

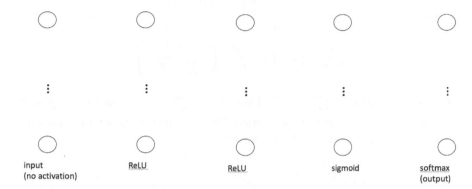

input
(no activation)

ReLU

ReLU

sigmoid

softmax
(output)

FIGURE 4.20: An Example Network for Discussing Backpropagation

Thus, if at a certain hidden layer we have ReLU activation, for an edge ij coming up to this $node_j$,

$$\triangle w_{ij} = \frac{\partial E}{\partial w_{ij}}$$
$$= \alpha o_i \delta_j$$

where

$$\delta_j = \frac{\partial E}{\partial o_j} \frac{\partial o_j}{\partial net_j}$$
$$= \left(\sum_{l \in L} \delta_l w_{jl} \right) \frac{\partial o_j}{\partial net_j}$$
$$= \left(\sum_{l \in L} \delta_l w_{jl} \right) r'(net_j)$$
$$= \begin{cases} \sum_{l \in L} \delta_l w_{jl} & \text{if } net_j > 0 \\ 0 & \text{otherwise.} \end{cases} \tag{4.28}$$

Thus, we find that the weight update computation is simpler than when using the sigmoid function. In modern ANNs, the ReLU activation is commonly used in hidden layers.

Backpropagation in the Softmax Layer: The softmax computation is given as

$$softmax\left(x^{(i)} \right) = s\left(x^{(i)} \right) = \frac{e^{x^{(i)}}}{\sum_{j=1}^{n} e^{x^{(j)}}}$$

for each $i = 1 \cdots n$. When we (partially) differentiate $s\left(x^{(i)} \right)$ w.r.t. $x^{(j)}$, we have two cases: i) when $i = j$, and ii) when $i \neq j$. We work out the two cases below.

When $i = j$, we have

$$\frac{\partial}{\partial x^{(i)}} s\left(x^{(i)}\right) = \frac{\partial}{\partial x^{(i)}} \left(\frac{e^{x^{(i)}}}{\sum_{j=1}^{n} e^{x^{(j)}}}\right)$$

Let $u = e^{x^{(i)}}$ and $v = \sum_{j=1}^{n} e^{x^{(j)}}$. Then, $u' = \frac{\partial u}{\partial x^{(i)}} = e^{x^{(i)}}$ and $v' = \frac{\partial v}{\partial x^{(i)}} = e^{x^{(i)}}$ as well since inside the summation for v, all terms except the one with index i are independent of $x^{(i)}$. Therefore,

$$\frac{\partial}{\partial x^{(i)}} s\left(x^{(i)}\right) = \frac{vu' - uv'}{v^2}$$

$$= \frac{e^{x^{(i)}} \sum_{j=1}^{n} e^{x^{(j)}} - \left(e^{x^{(i)}}\right)^2}{\left(\sum_{j=1}^{n} e^{x^{(j)}}\right)^2}$$

$$= \frac{e^{x^{(i)}} \left(\sum_{j=1}^{n} e^{x^{(j)}} - e^{x^{(i)}}\right)}{\left(\sum_{j=1}^{n} e^{x^{(j)}}\right)^2}$$

$$= s\left(x^{(i)}\right) \frac{\left(\sum_{j=1}^{n} e^{x^{(j)}} - e^{x^{(i)}}\right)}{\left(\sum_{j=1}^{n} e^{x^{(j)}}\right)}$$

$$= s\left(x^{(i)}\right) \left(1 - s\left(x^{(i)}\right)\right). \tag{4.29}$$

When $i \neq j$,

$$\frac{\partial}{\partial x^{(j)}} s\left(x^{(i)}\right) = \frac{\partial}{\partial x^{(j)}} \left(\frac{e^{x^{(i)}}}{\sum_{j=1}^{n} e^{x^{(j)}}}\right)$$

Let $u = e^{x^{(i)}}$ and $v = \sum_{j=1}^{n} e^{x^{(j)}}$, like before. Then, $u' = \frac{\partial u}{\partial x^{(j)}} = 0$ and $v' = \frac{\partial v}{\partial x^{(j)}} = e^{x^{(j)}}$. Therefore,

$$\frac{\partial}{\partial x^{(i)}} s\left(x^{(i)}\right) = \frac{vu' - uv'}{v^2}$$

$$= \frac{0 - e^{x^{(i)}} e^{x^{(j)}}}{\left(\sum_{j=1}^{n} e^{x^{(j)}}\right)^2}$$

$$= \frac{-e^{x^{(i)}} e^{x^{(j)}}}{\left(\sum_{j=1}^{n} e^{x^{(j)}}\right)^2}$$

$$= -s\left(x^{(i)}\right) s\left(x^{(j)}\right) \tag{4.30}$$

Therefore, we see that the derivatives can be computed easily using the input values to the softmax function $s(\ )$, without having to perform much additional computation.

At this time, let us look back at the ANNs we implemented in the program discussed in Section 4.4.2.5. These ANNs have a softmax layer for classification at the top, a sigmoid layer right below the softmax layer, one or more ReLU hidden layers, and the input layer at the bottom. We can use the derivatives of the softmax, sigmoid and ReLU activations as discussed in this section to implement backpropagation for such a network.

4.6   Loss Functions in Neural Networks

In our discussion of how neural networks are trained, we use an error function or a loss function. A loss or error function computes the error committed by a neural network in classifying a single example. In fact, it is a single number that represents the errors made by all of its neurons, which make computations based on the inputs coming to them and the weighting of each of these inputs. The input to the neural network depends on a training example that comes from a dataset, and thus, the input's components are fixed for a specific example. We also do not change the activation functions on any neuron in the entire network during training. The only things that can change are the weights in the network. That is why the weights are called parameters of the neural network. Since an ANN can have tens or hundreds of thousands, or even millions or billions of weights, the number of parameters in an ANN is enormous. Contrast this with the way we perform regression using standard approaches, possibly using regularization, when the number of parameters is approximately equal to the degree of the function we want to fit, and possibly a few more due to regularization. In tree-based classification and regression also, the number of parameters is usually quite small. Thus, ANNs present a much greater challenge because they attempt to solve a much larger optimization problem that is described by a very large number of parameters.

There is no way to solve such a complex optimization problem using any analytical technique. By this, we mean writing up a complex objective function that takes into account all the parameters, and then solving it by differentiating it and setting up the derivatives to 0, or some such technique. That is why neural networks use iterative techniques such as stochastic gradient descent (SGD) to perform optimization with a goal of obtaining an "optimal" set of weights. ANNs do not start by writing up an objective function that has all its parameters in it like we did in standard regression. An ANN starts with a simple objective function which is the loss or error function for an individual example. It backpropagates the effect of this error function backwards as it moves from the output layer back toward the input layer. As this computation moves backwards, it involves all the parameters or weights slowly and methodically, changing their values. It is a slow iterative process, but nonetheless if finds a way to change all the parameter values for each example it is shown and asked to learn from. This is done for each training example, and for a certain number of epochs. However, all this starts with a simple error or loss function computation at the output, for each example. What loss function we use affects how well or how quickly an ANN learns. The squared error loss is common in regression problems and can obviously be used in classification ANNs also, as we have already seen. However, researchers have come up with a number of other loss functions as well. In fact, squared error is not commonly used in modern classification ANNs although it is used for regression problems. A more common loss function for classification is what is called the cross-entropy loss. We have discussed entropy in the context of decision trees in Chapter 3. The same idea can be used to create a loss function for ANNs also. If we are performing binary classification, i.e., learning *yes* or *no* for certain classification task, we can use what is called (binary) cross-entropy, and for multi-class classification, we can use a generalization of it called the multi-class cross-entropy loss function.

Let us first understand cross-entropy loss. We have discussed the concept of entropy in Chapter 3 in the context of splitting a node in a decision tree considering the classes that examples in a mixed set belongs to. To recall, assume that we have a set S with n examples such that n_1 belong to class C_1 and the rest $n - n_1$ of the examples belong to class C_2. Then, the probability that a random example belongs to class C_1 is $p_1 = \frac{n_1}{n}$, and the probability that it belongs to class C_2 is $p_2 = 1 - p_1$. In such a situation, we say that the entropy of the set S is

$$entropy(S) = -p_1 \, ln \, p_1 - (1 - p_1) \, (1 - ln \, p_1). \qquad (4.31)$$

We used the concept of change in entropy to decide which feature should be used to split the tree. Entropy is used to compute the amount of mixing or so-called "chaos" in the set, considering the classes the examples belong to.

In this chapter, we use the idea of entropy in a slightly different way. Once again, suppose we have a set S of elements. Assume that we have two probability distributions associated with the elements in the set. The first discrete probability distribution tells us the probability that each element of the set belongs to a certain class C obtained one way, and the second also tells us the probability that each element belongs to the same class C, obtained in a different way.

In Figure 4.21, on the X-axis, we show the examples in the set, $x^{(i)}, i = 1 \cdots n$. On the Y-axis, we show the two discrete probability distributions. In particular $p_1^{(i)}$ shows the probability that $x^{(i)}$ belongs to class C obtained the first way, and $p_2^{(i)}$ shows the probability that $x^{(i)}$ belongs to class C obtained the second way, If now, we want to know how different the two discrete distributions are, we can compute what is called the cross-entropy between the

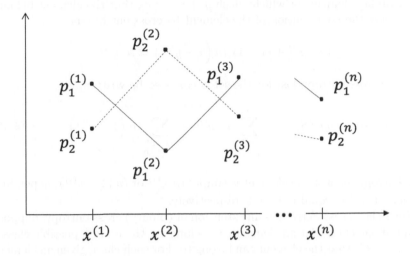

FIGURE 4.21: Two Discrete Probability Distributions

two discrete distributions, defined as given below.

$$cross\text{-}entropy(S) = -\sum_{i=1}^{N} \left[p_1^{(i)} \ ln \ p_2^{(i)} + \left(1 - p_1^{(i)} \right) \ ln \left(1 - p_2^{(i)} \right) \right] \quad (4.32)$$

In the context of ANNs, we have a training dataset containing n elements. Let the first discrete distribution come from the labels of a training dataset with binary classification. In other words, the $p_1^{(i)}$ values are simply the labels $y^{(i)}$ on element $x^{(i)}$,

$$p_1^{(i)} = y^{(i)}, \ i = 1 \cdots N. \quad (4.33)$$

The $p_2^{(i)}$ value for element $x^{(i)}$ is the probability obtained by the ANN classifiers that $x^{(i)}$ belongs to the class. In other words,

$$p_2^{(i)} = \hat{y}^{(i)}, \ i = 1 \cdots N. \quad (4.34)$$

Considering this, the cross-entropy formula becomes

$$cross\text{-}entropy(S) = -\sum_{i=1}^{N} \left[y^{(i)} \ ln \ \hat{y}^{(i)} + \left(1 - y^{(i)} \right) \ ln \left(1 - \hat{y}^{(i)} \right) \right] \quad (4.35)$$

The $y^{(i)}$ values are 0 or 1, depending upon if $x^{(i)}$ belongs to the class or not. The second discrete distribution corresponds to the probabilities that come out of an ANN after using a softmax layer.

For a training element for which $y^{(i)} = 0$, i.e., that the element does not belong to the class, the contribution of the element to cross-entropy is simply

$$0 \ ln \ \hat{y}^{(i)} + (1 - 0) \ ln \left(1 - \hat{y}^{(i)} \right) = ln \left(1 - \hat{y}^{(i)} \right).$$

For a training element for which which $y^{(i)} = 1$, i.e., that the element belongs to the class, the contribution of the element to cross-entropy is

$$1 \, ln \, \hat{y}^{(i)} + (1-1) \, ln \left(1 - \hat{y}^{(i)}\right) = ln \, \hat{y}^{(i)}.$$

Thus, the cross-entropy loss for the set S can also be written as

$$cross\text{-}entropy(S) = - \sum_{\substack{i=1 \\ |y^{(i)}=1}}^{N} ln \, \hat{y}^{(i)} - \sum_{\substack{i=1 \\ |y^{(i)}=0}}^{N} ln \left(1 - \hat{y}^{(i)}\right) \qquad (4.36)$$

Cross-entropy for a single element is simply $ln \, \hat{y}^{(i)}$ or $ln \left(1 - \hat{y}^{(i)}\right)$, depending on whether $y^{(i)}$ is equal to 0 or 1, respectively.

Multi-class cross-entropy is an extension of (binary) cross-entropy. Suppose instead of one class that an element can belong to, there are k possible classes $C_1, C_2, \cdots, C_k$ that the element can belong to. For each class, given an element $x^{(i)}$, we can obtain the binary cross-entropy. We add all binary cross-entropies to obtain the multi-class cross-entropy.

$$multi\text{-}class\text{-}cross\text{-}entropy(S) = \sum_{c=1}^{k} \sum_{i=1}^{N} \left(cross\text{-}entropy\left(\vec{x}^{(i)}\right)\right) \qquad (4.37)$$

where $cross\text{-}entropy\left(\vec{x}^{(i)}\right)$ is the contribution of $\vec{x}^{(i)}$ to the computation of total cross-entropy for the entire dataset.

Given a set of elements S, we can obtain cross-entropy loss for the entire set as described above. However, often we obtain the mean of the loss by dividing the individual losses by N, the number of elements in the set. We usually do this for all losses, including the squared error loss, discussed earlier.

4.7 Convolutional Neural Networks

In this section, we discuss a special kind of feedforward neural networks called Convolutional Neural Networks and show how they can be built with Keras.

Convolutions have been used in signal processing for a long time and have rich mathematical foundation. The field of Computer Vision, which attempts to detect objects and their features from images among other things, has used convolutions to aid in the process of detection. To describe what convolutional neural networks are, we first introduce the idea of convolutions in image processing. We do not go deep into the theory or use of convolutions, but get a feel for them so that we can understand convolutional neural networks.

4.7.1 Convolutions

4.7.1.1 Linear Convolutions

A convolution is a general purpose filter that can be used to extract features from an image as well as to change how an image looks. A convolution has a small matrix called a kernel that is applied to an image repeatedly using a mathematical operation. The mathematical operation for a single application of the kernel determines a new value of a central pixel by adding the weighted values of all its neighboring pixels. The output is a new modified filtered image. The kernel is placed on top of the bigger image and moved across the image well as down the image, covering the entirety of the image. Thus, a new "image" can be created from the old image. The new image captures or highlights some characteristics or features of the old image.

We start with a simple example in 1-D. Assume we have a vector of length 7 given to us. Assume all our vectors to be column vectors, and that we are given a vector $\vec{x}$ as shown below.

$$\vec{x}^T = \boxed{\begin{array}{|c|c|c|c|c|c|c|} \hline 1 & 5 & 6 & 1 & 2 & 4 & 3 \\ \hline \end{array}} \tag{4.38}$$

We are given another smaller vector, called a *convolution kernel* or simply *kernel*, $\vec{g}$.

$$\vec{g}^T = \boxed{\begin{array}{|c|c|c|} \hline \frac{1}{3} & \frac{1}{3} & \frac{1}{3} \\ \hline \end{array}} \tag{4.39}$$

A convolution kernel can be applied at various positions in the original vector, which we can call an image for now, although it is single-dimensional. For example, we can apply the convolution at location 2 of the image $\vec{x}$. Applying the convolution at a certain location in the image gives us a single value, which is obtained by placing $\vec{x}$ centered at location 2 below $\vec{x}$, multiplying the corresponding numbers in the two vectors and adding them up. Here we write the vectors in row form, i.e., we are writing their transposes.

$$
\begin{array}{|c|c|c|c|c|c|c|}
\hline
1 & 5 & 6 & 1 & 2 & 4 & 3 \\
\hline
\frac{1}{3} & \frac{1}{3} & \frac{1}{3} & \multicolumn{4}{c}{} \\
\hline
1 \times \frac{1}{3}+ & 5 \times \frac{1}{3}+ & 6 \times \frac{1}{3} & = 4 & & & \\
\hline
\end{array}
\tag{4.40}
$$

Thus, we can write

$$conv(\vec{x}, \vec{g}, 2) = 1 \times x[1] + 5 \times x[2] + 6 \times x[3] \tag{4.41}$$

$$= 1 \times \frac{1}{3} + 5 \times \frac{1}{3} + 6 \times \frac{1}{3}$$

$$= \frac{12}{3} = 4.$$

This convolutional kernel is simply computing the weighted average of the element at location 2 in $\vec{x}$ and its two neighbors.

In general, we can compute the convolution for every location in the vector $\vec{x}$. At location 1 and 7, we do not have two neighbors. In these locations, we

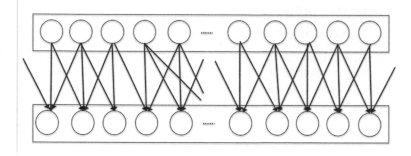

FIGURE 4.22: A Linear Convolutional Layer

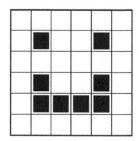

FIGURE 4.23: Black and White Pixels

can pad $\vec{x}$ on the left, and right, respectively, by 0, so that we can obtain an average of three numbers. For example, if we represent a word in a text in the form of a vector, as is commonly done, we may be able to combine the values of a few elements of this vector around a certain location in the vector using an averaging kernel like the one discussed here.

This idea of a linear convolution can be implemented in a neural network. Figure 4.22 shows that each node in the upper layer, which corresponds to a linear convolutional layer, is connected to only three nodes in the lower layer.

4.7.1.2   Two-Dimensional Convolutions

Textual documents often can be represented as vectors, although we do not discuss how this can be done in this book. However, images are not usually represented by vectors. An image is often represented by a two-dimensional matrix, where each cell in the matrix gives the intensity of a pixel, assuming it is black and white image. Grayscale images may use 8 bits to represent 256 levels of gray, 0 being black and 255 being white. Some grayscale representations like PNG and TIFF use 16 bits, with 0 being black and 65,536 being white. Consider the image given in 4.23. We can represent this image using 8 bits and 16 bits, as shown Table 4.5.

TABLE 4.6: Black and White
Representations of an Image

255	255	255	255	255	255
255	0	255	255	0	255
255	255	255	255	255	255
255	0	255	255	0	255
255	0	0	0	0	255
255	255	255	255	255	255

(a) 8-bit Representation

65536	65536	65536	65536	65536	65536
65536	0	65536	65536	0	65536
65536	65536	65536	65536	65536	65536
65536	0	65536	65536	0	65536
65536	0	0	0	0	65536
65536	65536	65536	65536	65536	65536

(b) 16-bit Representation

TABLE 4.7: Common Convolution Kernels for Image Processing

$$\begin{bmatrix} -1 & -1 & -1 \\ -1 & 8 & -1 \\ -1 & -1 & -1 \end{bmatrix} \quad \begin{bmatrix} 0 & -1 & 0 \\ -1 & 5 & -1 \\ 0 & -1 & 0 \end{bmatrix} \quad \frac{1}{9}\begin{bmatrix} 1 & 2 & 1 \\ 2 & 4 & 2 \\ 1 & 2 & 1 \end{bmatrix} \quad \frac{1}{9}\begin{bmatrix} 1 & 1 & 1 \\ 1 & 1 & 1 \\ 1 & 1 & 1 \end{bmatrix}$$

(a) Edge Detect   (b) Sharpen   (c) Gaussian Blur (d) Gaussian Blur

Given a grayscale image, we can convolve it for various effects such as smoothening, sharpening, intensifying, and embossing. It can also be used for tasks such as edge detection. Table 4.7 shows a few different convolutions used in image processing and computer vision.

Given an image and a convolution kernel, we can repeatedly apply the kernel to the image starting from the top left corner, moving slowly to the right. At the boundaries, we apply padding to repeat the numbers or wrap around. Once one application is over along the top of the image, we move the kernel down a bit and apply again across the image. We continue till we reach the right lower corner of the image. In other words, if the image is a matrix $\mathbf{X}$ and the convolution kernel is a matrix $\mathbf{K}$, then given an image pixel $X_{i,j}$, we can place the center of the kernel at the location i,j and then perform element by element multiplication and add up the numbers. We do so sweeping across the image from left to right, and then top to bottom. Figure 4.24 shows a grayscale image with an edge detection kernel overlaid in the center. The kernel's numbers are drawn in red and the area occupied by the kernel is shaded. The application of the kernel leads to a value of $0 \times (-1) + 0 \times$

255	255	255	255	255	0	255	255	255	255	255
255	255	255	255	0	0	0	255	255	255	255
255	255	255	0	0	255	255	0	255	255	255
255	255	255	0	255	255	255	255	0	255	255
255	255	0	0	0 (-1)	0 (-1)	0 (-1)	0	0	255	255
255	255	0	0	0 (-1)	0 (-8)	0 (-1)	0	0	255	255
0	0	255	255	255 (-1)	255 (-1)	255 (-1)	255	0	0	0
0	0	255	255	255	255	255	255	0	0	0
0	0	255	255	255	255	255	255	255	0	0
0	0	255	255	255	255	255	255	255	255	0
0	255	255	255	255	255	255	255	255	255	0

FIGURE 4.24: A simple grayscale image with a 3×3 edge detection kernel situated at location (6,6)

$$(-1) + 0 \times (-1) + 0 \times (-1) + 0 \times 8 + 255 \times (-1) + 255 \times (-1) + 255 \times (-1) = -765.$$

Assuming all values are positive, a negative number may be written as 0, if we were going to draw an image. In general, in a convolutional neural network, the negative number can remain as it is.

4.7.1.3   Three-Dimensional Convolutions

If we have a color image, the representation is usually in RGB format. This means that there are three so-called channels, one for red (R), one for green (G) and the third for blue (B). Each color has its own intensity value specified for a pixel. There are several representations for color images. A common representation requires 24 bits, wherein each channel has 8 bits. To represent an image in this form, we will need three 8-bit values at each pixel. We can think of the representation for an image being a 3-dimensional matrix, whose dimensions are $m \times n \times 3$, where m and n are the sizes in terms of pixels in the x and y directions. If we have an RGB representation for an image, the convolution must operate on all three channels as it is placed on a pixel.

4.7.2   Convolutional Neural Networks in Practice

A convolutional neural network (CNN) is a feedforward neural network with one or more convolutional layers like we have discussed in the previous section. A CNN usually has several convolutional layers. As alluded to earlier, a convolutional layer can be thought of as a feature extractor—a feature simply being something that stands out in an image. For example, the presence of a straight line at some location can be a useful feature. The presence of an

FIGURE 4.25: Examples of Features at Various Levels of Complexity

edge, a line that separates two parts of an image can also be a feature in computer vision. Some features are low-level whereas other features are high level. For example, the presence of a straight line of certain length at a certain slope at a certain location in an image can be considered a low-level feature. The presence of two lines of certain lengths meeting at a certain angle at a certain location is a slightly higher-level feature. The presence of a rectangle or a pentagon is a still higher-level feature. The presence of something that looks like an eye is a fairly high-level feature. The presence of two eyes is an even further higher-level feature. The presence of two eyes in a sketch with eyelashes is a further higher-level feature. Figure 4.25 shows some features that may be present in a sketch of a human face. The features are shown in the context of a simple sketch so as to explain the idea easily. In case of photographs or real images, even handwritten digits, the features are likely to be more complex. The point that is being made here is that an image processing or image classification program must be able to extract simple features from an image first, then extract more complex features out of the simple features, and continue to do so for several levels of sophistication. Once high level features have been extracted at a certain level, these features can be passed to a classifier to classify an image, e.g., to establish the fact that the image is that of a dog or cat, or that the image is the face of a particular person—in which case, we are solving the problem of face recognition.

Convolutional neural networks have been explicitly designed for processing images, like discussed in the previous paragraphs. The lowest level convolution extracts the lowest level features. These features are then passed on to another convolutional layer to extract higher level features out of the lowest level features. This process can be repeated a number of times by stacking a number of convolutional layers on top of each other. The sizes of the convolutional layers get progressively smaller, since fewer and fewer higher level features are extracted on higher levels of the stack. The number of convolutional layers depends on the designer of the network and the problem at hand. A deep neural network contains several such layers. In practice, the convolutional layers are interspersed with other types of layers such as pooling and dropout, which we do not discuss at length here; these layers are likely to further improve performance and generalization. Each convolutional layer usually has ReLU activation in all the neurons. On top of the convolutional layers, there are usually one or more sigmoid layers—for the purpose of actual classification. This portion is sometimes called the perceptron part of the CNN. The last layer of the perceptron has the classification results. However, modern classification ANNs have a softmax layer on top of the perceptron layer(s) so that the output of the entire CNN can be considered a vector of probabilities of belonging to the target classes. So, if there are ten target classes, the softmax layer will have ten neurons.

4.7.2.1 Convolutional Layer and Activation

The application of a convolution takes a matrix of values in an image and superimposes a filter at a certain location in the image, performs multiplications of the superimposed values, and then adds them up. This is repeated across and down the image by dragging the filter. Theoretically speaking, there is really no activation in a convolution unit. However, deep learning packages currently available allow us to add an activation function to neurons in a convolution layer. This means that the output of the convolution computation is given to an activation function. Usually, this function is ReLU. As a result, most discussion of CNNs conflate the convolution computation and ReLU computation into a single layer, usually called just a convolutional layer.

4.7.2.2 CNNs Using Keras and R

We can use Keras to develop CNN models and perform experiments. CNN models are feedforward models, and can be created like the models we have seen previously. Code to develop a CNN model for MNIST classification is given below.

```
1  #Make sure the libraries are installed. If not install them.
2  use_python("/usr/local/opt/python/bin/python3.6")
3  library (reticulate)
4  library (keras)
5
6  #get the MNIST dataset
7  mnist <- dataset_mnist()
8
9  #Set up training and testing subsets
10 x_train <- mnist$train$x
11 y_train <- mnist$train$y
12
13 x_test <- mnist$test$x
14 y_test <- mnist$test$y
15
16 #Reshape and pre-process the input
17 x_train <- array_reshape(x_train, c(nrow(x_train), 28, 28, 1))
18 x_test <- array_reshape(x_test, c(nrow(x_test), 28, 28, 1))
19 x_train <- x_train / 255
20 x_test <- x_test / 255
21 y_train <- to_categorical(y_train, 10)
22 y_test <- to_categorical(y_test, 10)
23
24 #Set up the neural model
25 modelCNN1 <- keras_model_sequential ()
26 modelCNN1 <- layer_conv_2d (modelCNN1, filters = 32,
27    kernel_size = c(3,3),activation = 'relu', input_shape=c(28,28,1))
28 modelCNN1 <- layer_conv_2d (modelCNN1, filters = 64,
```

```
29        kernel_size = c(3,3), activation = 'relu')
30  modelCNN1 <- layer_conv_2d (modelCNN1, filters = 64,
31        kernel_size = c(3,3), activation = 'relu')
32  modelCNN1 <- layer_flatten (modelCNN1)
33  modelCNN1 <- layer_dense (modelCNN1, units = 64, activation = 'sigmoid')
34  modelCNN1 <- layer_dense (modelCNN1, units = 10, activation = 'softmax')
35
36  #print a summary of the model
37  summary(modelCNN1)
38
39  #Compile the model, i.e., set it up to run
40  compile(modelCNN1, loss = 'categorical_crossentropy', optimizer = optimizer_rmsprop(),
41        metrics = c('accuracy'))
42
43  #the fit function runs the model. We save the run history in a variable
44  history <- fit(modelCNN1, x_train, y_train,epochs = 10, batch_size = 128,
45        validation_split = 0.2)
46
47  #plot how training and validation went
48  plot (history)
49
50  #evaluate the model's performance on test data
51  evaluate(modelCNN1, x_test, y_test)
```

Lines 1-15 are exactly the same as explained earlier. In these lines, we make sure we have a correct version of Python (this line may not be necessary depending on configuration), load libraries, load the MNIST dataset and obtain four subsets of data: training examples, training labels, testing examples and testing labels. Lines 17-18 reshape the training example tensors as expected by a CNN. A CNN expects the training examples to be given like an image, unlike the simple feedforward network we discussed earlier; in the earlier programs, each image of the digits, although given as image in 2-D originally, was converted to an 1-D vector. But, in the case of CNNs in Keras, the image is usually converted to a 3-D image. The digit images are black and white. Therefore, they have only one so-called channel. If they were in color, they would have three channels, corresponding to Red (R), Green (G) and Blue (B). Each original digit image is 28 × 28 pixels. To input the image to a CNN, we imagine it to be a 3-D image of dimensions 28 × 28 × 1. Thus, the 2-D matrix is reshaped into a 3-D matrix. A general name for matrices is tensors. In terms of formal ANN literature, we reshape 2-D tensors in the training examples to 3-D tensors. Lines 19-22 are exactly the same as in the program discussed earlier in Section 4.4.2.5. We convert the pixel values in the example images to real values between 0 and 1 instead of integer values between 0 and 255. We also convert the y values in the training and testing datasets to 1-hot vectors, each vector being 10 long.

Lines 25-34 set up the CNN model and represent the heart of the program. Like what we saw in the program in Section 4.4.2.5, they are easy to follow. Line 25 starts to build a sequential or feedforward model. Lines 26-27 build the first convolutional layer. Since there is only one channel in the image, we build a so-called 2-D convolutional layer. There are 32 different convolutions. In other words, the CNN is going to learn the weights for 32 different local filters or convolutions; each filter looks at a small 3 × 3 region of the image at

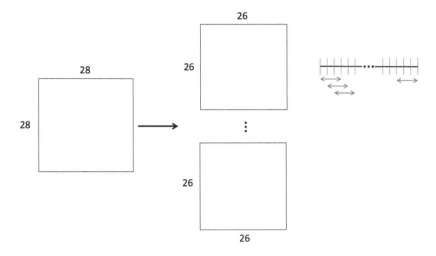

FIGURE 4.26: 2-D Convolutions and Strides

a time, and is moved over the entire image. A 2-D convolution runs down the image in 2-D, i.e., horizontally and vertically. The default assumption is that there is no padding. When there is no padding, depending on the dimensions of the image and the dimensions of the filter, the dimensions of the tensors may reduce automatically after a filter is applied on the entire image. For each filter, the CNN learns the weights. Since each filter is 3×3, there are 9 weights associated with it. There are 32 filters; as a result, the connection between the input layer and the first convolution layer has $32 \times 9 = 288$ weights or parameters to be learned as far as convolutions go. Each convolution may have an extra input called bias, which is set to 1. Biases are not necessary in deep networks, and the default presence is due to historical reasons. The presence of an extra weight in each convolution or filter make the total number of parameters between the input and the first convolutional layer $288 + 32 = 320$.

Consider one of the dimensions of the input, say X. If we have a ruler that is 28 long, and we place a stick 3 units long on the extreme left, and move it to the right by 1 spot, and continue to do so until the stick is at the extreme right of the ruler, we have 26 different placements of the stick. Since a filter is 3×3 in size and is moved along Y as well, it can be placed on a 28×28 image 26×26 times. Each placement of the filter is accompanied by a dot product based computation, and the entire process results in 26×26 new values being produced. This is shown schematically in Figure 4.26. Since there are 32 of these filters, the output of the applications of the 32 filters transforms a $32 \times 32 \times 1$ tensor of values to a $26 \times 26 \times 32$ tensor of values. Since each tensor is 3-D, it can be thought of as a volume, and the application of the filters can be thought of one volume being converted to another. In fact, the transformation between each pair of consecutive layers in the convolutional part (not the dense perception part) can be thought of as volume to volume (or, tensor to tensor) transformation.

The shapes of the volumes or tensors as we move from input to the output of the network can be used to describe its architecture. Keras provides us with such a description if we use the **summary** command after the network has been built partially or fully.

```
> summary(modelCNN1)
Model: "sequential_17"
```

Layer (type)	Output Shape	Param #
conv2d_46 (Conv2D)	(None, 26, 26, 32)	320
conv2d_47 (Conv2D)	(None, 24, 24, 64)	18496
conv2d_48 (Conv2D)	(None, 22, 22, 64)	36928
flatten_15 (Flatten)	(None, 30976)	0
dense_31 (Dense)	(None, 64)	1982528
dense_32 (Dense)	(None, 10)	650

```
Total params: 2,038,922
Trainable params: 2,038,922
Non-trainable params: 0
```

The summary above gives the architecture of the CNN we have built. We have three 2-D convolutional layers. The first convolutional layer produces a tensor of shape $26 \times 26 \times 32$. This means we have 32 slices, each of size 26×26. Although, nominally we say that each filter is of size 3×3, in reality each filter applied to all 32 layers at once. Thus, the actual filter size is $3 \times 3 \times 32 = 288$. A single 32-deep filter is of size 288, and we have 64 such filters, leading to a total of $288 \times 64 = 36,864$ parameters or weights needed to connect the two layers. If we add a bias input for each filter, the total number of parameters becomes $36,864 + 64 = 36,928$. This is shown in Figure 4.27.

When the last convolutional layer is flattened, it has a single row of $22 \times 22 \times 64 = 30,976$ units. The next layer is 1-D with 64 units. As a result, there are $30,976 \times 64 = 1,982,464$ weights. Each of the 64 units has a bias input as well, making a total of $1,982,528$ weights or parameters. Connecting 64 nodes densely to a layer with ten nodes requires 640 parameters, to which we add 10 bias weights to get 650. The total number of parameters in our ANN is quite large, at over 2 million.

We compile the model and run it like the program in Section 4.4.2.5. Since the number of parameters is large, the processing becomes significantly slower. The progress of training is given in Figure 4.28 in terms of the accuracy metric. It starts with a validation accuracy of 0.9818 in epoch 1 and goes up to validation accuracy of 0.9892 at the end of epoch 10 in a particular run.

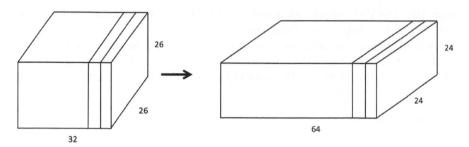

FIGURE 4.27: Transformation of Tensor Volumes in a Neural Network from Layer to Layer

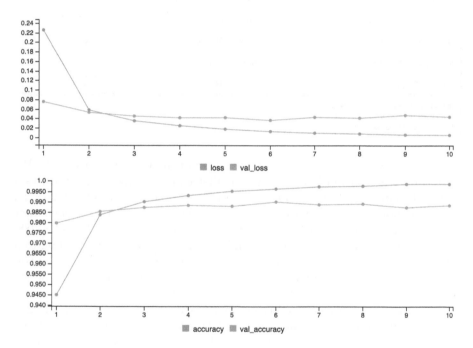

FIGURE 4.28: Loss and Accuracy Metrics for CNN with Three Convolutional Layers (No Pooling)

4.7.3 Pooling Layers in CNNs

A pooling layer is somewhat like a convolutional layer in that it operates on a small region of the feature space extracted by a convolutional layer, but instead of multiplying each element in the image (feature space) and adding them up, a pooling layer may pick up the largest element in a small region. This is called *max pooling*. Like convolutions, the pooling area can be variable, but always a square. So, if the size is 4×4, it picks up the largest element from

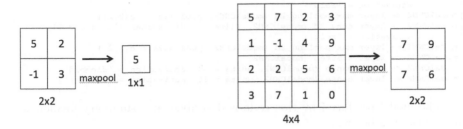

FIGURE 4.29: Maxpooling in CNN

the image (feature space) when superimposed. If the size of the pooling filter is 2×2, it picks up the largest element in a 2×2 area of the feature map. If we use a stride of 2 with a pooling size of 2×2, the feature map is essentially reduced by a size of 2. As a result, pooling is also called downsampling. Figure 4.29 shows the use of a 2×2 max pooling filter in one location in the feature space.

The purpose of downsampling using a pooling layer is to make the CNN a little more robust in feature extraction. A convolutional layer extracts features from an input layer or a layer of lower level features. The convolutional layer is quite precise in picking up locations of features. For example, if we are trying to pick up a vertical line at a certain location as a feature, the line picked up may not be exactly vertical, with some feature elements picked up being not properly aligned, but if we allow for small misalignments, say within 2 horizontal pixels, we may still be able to extract the vertical line as a feature. Max pooling allows the CNN to do so. There is another type of pooling that is commonly used also. It is called average pooling. Average pooling simply averages the numbers within the matrix to obtain an average estimation of the feature values within a small region.

Figure 4.29 shows also the downsampling performed by a 2×2 maxpooler with a stride of 2. In general, such maxpooling reduces the size of an image or feature space to half size in both directions for a 2-D setup.

4.7.3.1 CNN using Pooling in R

We can add pooling layers in various locations in a CNN using Keras. The code changes only minimally. We use maxpooling after each of the three convolutional layers in the code below. There is no rule that we should use pooling after every convolutional layer, but that is what we do in this example. This practice is common. Pooling is not usually used after dense layers.

```
#Set up the neural model
modelCNN2 <- keras_model_sequential ()
modelCNN2 <- layer_conv_2d (modelCNN2, filters = 32, kernel_size = c(3,3),
       activation = 'relu', input_shape=c(28,28,1))
modelCNN2 <- layer_max_pooling_2d (modelCNN2, pool_size = c(2,2))
modelCNN2 <- layer_conv_2d (modelCNN2, filters = 64, kernel_size = c(3,3),
```

```
        activation = 'relu')
modelCNN2 <- layer_max_pooling_2d (modelCNN2, pool_size = c(2,2))
modelCNN2 <- layer_conv_2d (modelCNN2, filters = 64, kernel_size = c(3,3),
        activation = 'relu')
modelCNN2 <- layer_max_pooling_2d (modelCNN2, pool_size = c(2,2))
modelCNN2 <- layer_flatten (modelCNN2)
modelCNN2 <- layer_dense (modelCNN2, units = 64, activation = 'sigmoid')
modelCNN2 <- layer_dense (modelCNN2, units = 10, activation = 'softmax')
```

The model produced by this code is described in a summary fashion by Keras as given below.

```
> summary(modelCNN2)
Model: "sequential_18"
```

Layer (type)	Output Shape	Param #
conv2d_49 (Conv2D)	(None, 26, 26, 32)	320
max_pooling2d_28 (MaxPooling2D)	(None, 13, 13, 32)	0
conv2d_50 (Conv2D)	(None, 11, 11, 64)	18496
max_pooling2d_29 (MaxPooling2D)	(None, 5, 5, 64)	0
conv2d_51 (Conv2D)	(None, 3, 3, 64)	36928
max_pooling2d_30 (MaxPooling2D)	(None, 1, 1, 64)	0
flatten_16 (Flatten)	(None, 64)	0
dense_33 (Dense)	(None, 64)	4160
dense_34 (Dense)	(None, 10)	650

```
Total params: 60,554
Trainable params: 60,554
Non-trainable params: 0
```

We see that the model has only $60,554$ trainable parameters compared to 2.038 million parameters without pooling. This is a stunning reduction of 97% in the size of the ANN, making it vastly more efficient. However, performance in terms of accuracy actually does not suffer. The validation accuracy started at 0.9282 in epoch 1 in a particular run and went up to 0.9837 at the end of epoch 10. Thus, pooling is an extremely important tool in building CNN architectures that work.

4.7.4   Regularization in CNNs

Neural networks usually have a lot of parameters and as a result, it is difficult to find the optimal weights. ANNs, like all supervised machine learning techniques, may overfit to the training data. It means that the ANN memorizes some or a lot of the training data, gets high performance metrics on the training data, but does not produce good results on unseen data. In other words, the ANN is not able to generalize well, which is the ultimate aim of any machine learning approach. A machine learning algorithm is supposed to learn from the training data, but generalize to the extent possible so that it works well with unseen data. In other words, it should not learn all patterns in the training data, just the most significant ones in some sense, that are likely to be present in unseen data as well. Generalization requires that the trained model performs well on unseen data. It is something difficult to guarantee, and in ANNs, there are several approaches that have been proposed.

One way to reduce the amount of overfitting is by regularization, as seen in Section 2.6, during our discussion on regression. Similar regularization, e.g., L2 regularization, can be used in ANNs also, although we do not discuss it in this book. Two other methods to perform regularization have also been found useful in ANNs: *dropout* and *batch normalization*.

4.7.4.1 Dropout for ANN Regularization

Dropout essentially means that during training of an ANN, a certain percentage of the nodes are assumed to not exist. That is, all the weights to them are set to 0. The weights are not used in forward computation and not updated during backpropagation. The dropped nodes are randomly picked. The motivation is that when we randomly remove certain nodes, the other nodes in the layer have to work together better to learn the essential patterns. Since the number of nodes becomes lower, the potential to learn non-essential patterns is reduced, because the ANN has to work with less. The hypothesis is that this leads to better generalization. The idea of dropout was proposed by one of the biggest names in ANN and was popular a few years ago. However, researchers have shown recently that the effect of dropout is unclear and controversial.

Potentially dropout can be used at any layer, although it was used at the dense layers when proposed. Dropout is used during training only. During testing or application of a trained network, no nodes are dropped and the entire network is used.

Dropout in Keras: Although dropout randomly removes neurons from a hidden layer, in terms of programming statements, it is usually added as if it is a layer by itself. The code below shows the use of 50% dropout in the dense layer before the softmax layer.

```
modelCNN2 <- keras_model_sequential ()
modelCNN2 <- layer_conv_2d (modelCNN2, filters = 32, kernel_size = c(3,3),
    activation = 'relu', input_shape=c(28,28,1))
modelCNN2 <- layer_max_pooling_2d (modelCNN2, pool_size = c(2,2))
modelCNN2 <- layer_conv_2d (modelCNN2, filters = 64, kernel_size = c(3,3),
    activation = 'relu')
modelCNN2 <- layer_max_pooling_2d (modelCNN2, pool_size = c(2,2))
modelCNN2 <- layer_conv_2d (modelCNN2, filters = 64, kernel_size = c(3,3),
    activation = 'relu')
modelCNN2 <- layer_flatten (modelCNN2)
modelCNN2 <- layer_dense (modelCNN2, units = 64, activation = 'sigmoid')
modelCNN2 <- layer_dropout(modelCNN2, rate = 0.5)
modelCNN2 <- layer_dense (modelCNN2, units = 10, activation = 'softmax')
```

The use of dropout is shown in second line from the end. The use of even a single dropout layer makes the ANN slower. In a run of this modified ANN, it produces a validation accuracy of 0.9896, about the same as without the dropout layer. However, it can also be because the model without dropout works so well that it is difficult to improve much further.

To check if using dropout at more layers is beneficial, we insert dropout after every convolutional layer as well as the sigmoid layer.

```
modelCNN2 <- keras_model_sequential ()
modelCNN2 <- layer_conv_2d (modelCNN2, filters = 32, kernel_size = c(3,3),
                     activation = 'relu', input_shape=c(28,28,1))
modelCNN2 <- layer_dropout(modelCNN2, rate = 0.5)
modelCNN2 <- layer_max_pooling_2d (modelCNN2, pool_size = c(2,2))
modelCNN2 <- layer_conv_2d (modelCNN2, filters = 64, kernel_size = c(3,3),
                     activation = 'relu')
modelCNN2 <- layer_dropout(modelCNN2, rate = 0.5)
modelCNN2 <- layer_max_pooling_2d (modelCNN2, pool_size = c(2,2))
modelCNN2 <- layer_conv_2d (modelCNN2, filters = 64, kernel_size = c(3,3),
                     activation = 'relu')
modelCNN2 <- layer_dropout(modelCNN2, rate = 0.5)
modelCNN2 <- layer_flatten (modelCNN2)
modelCNN2 <- layer_dense (modelCNN2, units = 64, activation = 'sigmoid')
modelCNN2 <- layer_dropout(modelCNN2, rate = 0.5)
modelCNN2 <- layer_dense (modelCNN2, units = 10, activation = 'softmax')
```

The use of 4 dropout layers makes training considerably slower. However, the overall results do not change much. Note that optimization is performed using stochastic approaches, and as a result, the results may come out different in different runs. After 10 epochs, the validation accuracy comes out as 0.9896, in a run of the program. However, if we compare the way loss and accuracy change over the epochs (see Figure 4.30), we see that the use of multiple dropouts makes the difference between the training and validation losses (and accuracies) much smaller compared to the case when only one dropout was used. This shows that, at least in some cases, the extensive use of dropouts reduces the amount of overfitting in neural networks. However, this is not

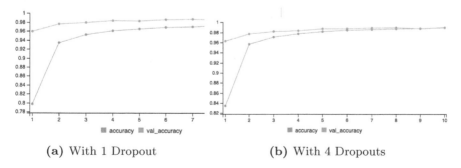

(a) With 1 Dropout   (b) With 4 Dropouts

FIGURE 4.30: Training and Validation Accuracies with Dropouts

likely to be true always, i.e., the benefits accrued from the use of dropouts are not always clear. The issue remains unsettled in current literature.

4.7.4.2 Batch Normalization for Regularization

We have already showed that normalization of the input improves performance of a network. In particular, when classifying the MNIST dataset, dividing the pixel values (each between $0 - 255$) by 255, thus normalizing them to be real numbers between 0 and 1, improves accuracy of classification substantially. One reason for this is likely to be that the weight values are usually randomly initialized to be small positive and negative values centered around 0, and having input values in the same range improves performance, since bigger values do not completely overwhelm smaller values. In addition, inputs are processed by ANNs in small batches, say of size 64 or 128 each. These are called *mini-batches* or simply batches. All the elements in a batch are processed and calculations are performed, but values are not updated by back propagation till the end of a batch. The activation and gradient values are computed for each example in a batch; the averages of the activation and gradient values are computed and are used in the computations necessary for backpropagation.

Batch normalization, which was introduced in 2015, goes a little beyond dividing all input numbers by the difference between the highest value and the lowest value, to bring all values to the range $0 - 1$. It is based on the idea of z-normalization. Z-normalization replaces a number in a set of numbers by its "standard score" or its value if the standard deviation of the set of numbers is taken as the unit. A number $x^{(i)}$ in the set is replaced by $z^{(i)}$ such that

$$z^{(i)} = \frac{x^{(i)} - m}{s} \tag{4.42}$$

where m is the mean of the numbers and s is the standard deviation of the set of numbers, with

$$m = \frac{1}{b} \sum_{i=1}^{b} x^{(i)}, \text{ and}$$

$$s = \frac{1}{b} \sqrt{\sum_{=1}^{b} (x^{(i)} - m)^2}, \tag{4.43}$$

where b is the number of numbers in the set. In batch normalization, we perform a similar normalization for each dimension of the input. Assume the input has n dimensions, i.e.,

$$\vec{x}^{(i)} = \left\langle x_1^{(i)}, x_2^{(i)} \cdots x_n^{(i)} \right\rangle \tag{4.44}$$

where $\vec{x}^{(i)}$ is the ith input. The program looks at all values for each dimension, $k = 1 \cdots n$. For each dimension k, it performs normalization considering b

inputs in a batch. In particular, the normalization performed computes the following,

$$m_k = \frac{1}{b} \sum_{j=1}^{b} x_k^{(j)}, \text{ and}$$

$$s_k = \frac{1}{b} \sum_{j=1}^{b} \left(x_k^{(j)} - m_k \right)^2 \tag{4.45}$$

to compute mean and standard deviation for a dimension k. Then, each value in dimension k is normalized as

$$z_k^{(j)} = \frac{x_k^{(j)} - m_k}{\sqrt{s_k^2 + \epsilon}}, \ k = 1 \cdots n, \ j = 1 \cdots b \tag{4.46}$$

where ϵ is a small number added to s_k^2 to ensure that there is no division by 0. However, $z_k^{(j)}$ is modified further to obtain a value $u_k^{(j)}$ where

$$u_k^{(j)} = \gamma_k \, z_k^{(j)} + \beta_k \tag{4.47}$$

where γ_k and β_k are two parameters that are learned during the optimization process, for each input dimension k. Thus, batch normalization is a function bn such that

$$bn\left(x^{(i)} \right) = u^{(i)} = \left\langle u_1^{(i)}, u_2^{(i)} \cdots u_n^{(i)} \right\rangle \tag{4.48}$$

with parameters γ_k, β_k, $k = 1 \cdots n$, two parameters for each dimension of the input. $\vec{u}^{(i)}$ is passed onto the next layer of the neural network.

We have discussed how batch normalization is performed during training. It can be shown that the bn function is differentiable, and backpropagation can be altered to take it into account, although we do not discuss the details here.

In fact, batch normalization can be performed in any layer of a neural network, not just the input layer. Batch normalization makes the neural network slower, but it has been shown that it makes the network overfit less, i.,e., generalize better.

Batch Normalization in Keras: The use of batch normalization layer on the dense layer before the softmax layer produces a validation accuracy of 0.9902 in a run of the network after ten epochs. Batch normalization also makes the processing slower compared to not using it; in our experiments, the time is comparable to using one dropout layer for one batch normalization. The use of four batch normalizations, as shown below, makes the training quite a bit slower.

```
modelCNN2 <- keras_model_sequential ()
modelCNN2 <- layer_conv_2d (modelCNN2, filters = 32, kernel_size = c(3,3),
                    activation = 'relu', input_shape=c(28,28,1))
modelCNN2 <- layer_batch_normalization (modelCNN2)
```

```
modelCNN2 <- layer_max_pooling_2d (modelCNN2, pool_size = c(2,2))
modelCNN2 <- layer_conv_2d (modelCNN2, filters = 64, kernel_size = c(3,3),
                           activation = 'relu')
modelCNN2 <- layer_batch_normalization (modelCNN2)
modelCNN2 <- layer_max_pooling_2d (modelCNN2, pool_size = c(2,2))
modelCNN2 <- layer_conv_2d (modelCNN2, filters = 64, kernel_size = c(3,3),
                           \UTF{00DF}activation = 'relu')
modelCNN2 <- layer_batch_normalization (modelCNN2)
modelCNN2 <- layer_flatten (modelCNN2)
modelCNN2 <- layer_dense (modelCNN2, units = 64, activation = 'sigmoid')
modelCNN2 <- layer_batch_normalization (modelCNN2)
modelCNN2 <- layer_dense (modelCNN2, units = 10, activation = 'softmax')
```

The validation accuracy after 10 epochs in one run comes out to be 0.9886, comparable to the result with one batch normalization. The accuracy graphs for the two cases are shown in Figure 4.31.

The initial understanding was that batch normalization improves the performance of a neural network by reducing the shift in the distribution of inputs to a layer of the network. The input to a layer of an ANN depends on the output of the previous layer. The outputs of a layer for a batch have a certain probability distribution. Assuming the distribution is Gaussian, it is a bell-shaped curve. Over time, the location of the top of the bell or the shape of the tails may change. Changes in lower layers, even small changes, may affect upper layers substantially, affecting backpropagation even more. The use of batch normalization resets the distribution in some sense, reducing the effect of distribution shift, if any. However, as of the writing of this book, there is substantial debate on this explanation of batch normalization, although it seems to make deep networks learn better.

At this time (mid-2022), both dropout and batch normalization are used by practitioners of deep learning, although there is a movement toward preferring batch normalization.

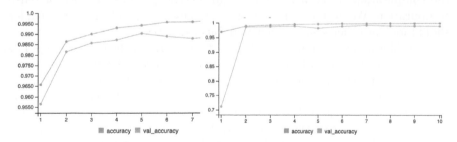

(a) With 1 Batch Normalization   (b) With 4 Batch Normalizations

FIGURE 4.31: Training and Validation Accuracies with Dropouts

4.8 Matrix or Tensor Formulation of Neural Networks

As we have seen, even simple neural networks that are deep can have a very large number of parameters. Millions or even billions of parameters are not uncommon. One reason deep neural networks have become useful, effective and popular is because they are usually implemented in GPUs (Graphical Processing Units). GPUs are special-purpose computing hardware that were initially designed to perform expensive computations needed for computer graphics, animation, special effects creation in movies, and game programming. These areas require computing almost exclusively with large vectors and matrices, and also need these computations to be very efficient so that movements of realistic images and characters look almost life-like.

To be able to implement neural networks on GPUs, we have to be able to express the computations performed in terms of vectors and matrices. Vectors are one-dimensional, whereas matrices can have any number of dimensions. Generally, matrices in 2-, 3- or 4-dimensions are used to represent gray images, color images and moving images or videos, respectively. A general name for numeric data structures of various dimensions is *tensors*. So, we can think of vectors as 1-D tensors, gray scale images as 2-D tensors, color images as 3-D tensors and videos as 4-D tensors. Thus, depending on the type of input a neural network works on, the input can be a 1- to 4-D tensor. The computations performed by the neural network, such as weighted summation of inputs and application of activation functions, pooling and convolutions, have to be expressed as tensor computations as well, so that they can be implemented efficiently on GPUs. As a result, the entire computation performed by a neural network can be thought of as a sequence of tensor computations.

In this section, we look at a simple feedforward neural network and see how the computations from layer to layer can be thought of as matrix or tensor computation. We do not work out any complex cases, and interested readers should look at other sources such as published papers.

Consider a neural network with only two layers, as shown in Figure 4.32. It is a dense feedforward network with the two layers labeled $l-1$ and l. The input coming from the nodes in layer $l-1$ are numbered $x_1^{l-1}, \cdots, x_{n_{l-1}}^{l-1}$, where the superscript is the layer number and the subscript is the node number within the layer. In layer l, we see the net input to a node. It is the weighted sum of the inputs from the prior layer. The net inputs to the nodes in layer l are denoted $net_1^{l-1}, \cdots, net_{n_{l-1}}^{l-1}$, where the superscript is the layer number and n_l is the number of nodes in layer l. The activation function at a node is labeled f^l; it is the same for all nodes in a layer. The output of node i in layers l is labeled o_i^l. This o_i^l is actually x_i^l, which is used as input to layer $l+1$. The weight on the edge between node j in layer $l-1$ and node i is layer l is given as $w_{j,i}^{l-1,l}$. To keep our notations simple, we will not write the

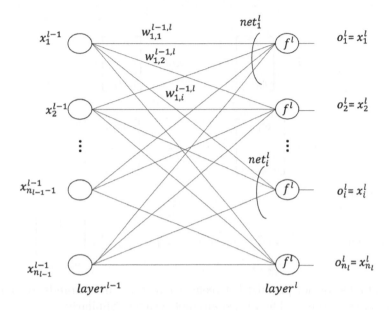

FIGURE 4.32: Two feedforward layers of a neural network

superscripts that signify the layer number. So, instead of x_i^{l-1}, $w_{j,i}^{l-1,l}$ net_i^l and o_i^l, we write x_i, $w_{j,i}$ net_i and o_i, respectively.

The net input to the nodes in layer l can be written as follows. Note that there may be a bias input of magnitude 1 to each non-input layer node; it is not shown here.

$$net_1 = w_{1,1}x_1 + \cdots + w_{n_{l-1},1}x_{n_{l-1}}$$

$$\vdots$$

$$net_i = w_{1,i}x_1 + \cdots + w_{n_{l-1},i}x_{n_{l-1}}$$

$$\vdots$$

$$net_{n_l} = w_{1,n_l}x_1 + \cdots + w_{n_{l-1},n_l}x_{n_{l-1}}$$

$$(4.49)$$

We now formulate the computations performed in terms of tensors—vectors and matrices. Let us define some vectors first: $\vec{x}, \overrightarrow{net}$ and $\overrightarrow{w_1}, \cdots, \overrightarrow{w_{n_l}}$ as follows.

$$\vec{x} = \begin{bmatrix} x_1 \\ \vdots \\ x_{n_{l-1}} \end{bmatrix} \qquad \overrightarrow{net} = \begin{bmatrix} net_1 \\ \vdots \\ net_{n_l} \end{bmatrix}$$

$$\overrightarrow{w_1} = \begin{bmatrix} w_{1,1} \\ \vdots \\ w_{n_{l-1},1} \end{bmatrix} \quad \cdots \quad \overrightarrow{w_{n_l}} = \begin{bmatrix} w_{1,n_l} \\ \vdots \\ w_{n_{l-1},n_l} \end{bmatrix} \qquad (4.50)$$

We can write

$$net_1 = \overrightarrow{w_1} \circ \vec{x} = \vec{x}^T \cdot \overrightarrow{w_1} = \overrightarrow{w_1}^T \cdot \vec{x} = \begin{bmatrix} x_1 \cdots x_{n_{l-1}} \end{bmatrix} \cdot \begin{bmatrix} w_{1,1} \\ \vdots \\ w_{n_{l-1},1} \end{bmatrix} \qquad (4.51)$$

Here, $\circ$ stands for dot product between two vectors, and $\cdot$ stands for normal product of two vectors. The $\cdot$ is usually not written. Similarly,

$$net_2 = \overrightarrow{w_2} \circ \vec{x} = \vec{x}^T \cdot \overrightarrow{w_2} = \overrightarrow{w_2}^T \cdot \vec{x}$$

and

$$net_{n_l} = \overrightarrow{w_{n_l}} \circ \vec{x} = \vec{x}^T \cdot \overrightarrow{w_{n_l}} = \overrightarrow{w_{n_l}}^T \cdot \vec{x}.$$

Let us write all net_is as a vector

$$\overrightarrow{net} = \begin{bmatrix} net_1 \\ \vdots \\ net_{n_l} \end{bmatrix}.$$

Each net_i is a dot product of two vectors, or alternatively a regular vector product of a transposed vector and another vector. We can write the following.

$$\overrightarrow{net} = \begin{bmatrix} net_1 \\ \vdots \\ net_{n_l} \end{bmatrix} = \begin{bmatrix} \overrightarrow{w_1} \circ \vec{x} \\ \vdots \\ \overrightarrow{w_{n_l}} \circ \vec{x} \end{bmatrix} = \begin{bmatrix} w_{1,1} & \cdots & w_{1,n_{l-1}} \\ \vdots & \cdots & \vdots \\ w_{n_l,1} & \cdots & w_{n_l,n_{l-1}} \end{bmatrix} \begin{bmatrix} x_1 \\ \vdots \\ x_{n_{l-1}} \end{bmatrix} = \mathbf{W}\,\vec{x}.$$

$$(4.52)$$

Here,

$$\mathbf{W} = \begin{bmatrix} w_{1,1} & \cdots & w_{1,n_{l-1}} \\ \vdots & \cdots & \vdots \\ w_{n_l,1} & \cdots & w_{n_l,n_{l-1}} \end{bmatrix}$$

where each column represents the weights on the edges incident on an output node. Thus, all the outputs can be obtained by a matrix-vector or tensor product that a GPU can compute efficiently.

4.8.1   Expressing Activations in Layer l

In layer l in Figure 4.32, every node has an activation function that acts upon the net input coming to the node. The activation function is the same in all nodes in a layer. Let us call it $f(\ )$. Thus, we can write

$$o_1 = f\left(\vec{w_1} \circ \vec{x}\right)$$
$$o_2 = f\left(\vec{w_2} \circ \vec{x}\right)$$
$$\vdots$$
$$o_{n_l} = f\left(\vec{w_{n_l}} \circ \vec{x}\right).$$

If we express all outputs as a vector

$$\vec{o} = \begin{bmatrix} o_1 \\ \vdots \\ o_{n_l} \end{bmatrix},$$

we can write

$$\vec{o} = \begin{bmatrix} f\left(\vec{w_1} \circ \vec{x}\right) \\ f\left(\vec{w_2} \circ \vec{x}\right) \\ \vdots \\ f\left(\vec{w_{n_l}} \circ \vec{x}\right) \end{bmatrix} = \mathbf{f}\left(\mathbf{W}\,\vec{x}\right) \tag{4.53}$$

where we make the assumption that $\mathbf{f}(\ )$ represents the application of function $f(\ )$ to every element of a vector to produce a vector output.

4.8.2   Generalizing to a Bigger Network

Consider a bigger network of L layers, with the first layer numbered 0, and the last layer numbered $L-1$, as shown in Figure 4.33. Let

$$\mathbf{W}^l = \left[w_{jk}^{l-1,l}\right]$$

be the weights between layers $l-1$ and l, where $w_{jk}^{l-1,l}$ is the weight between the jth node of layer $l-1$ and the kth node of layer l. Then, the overall network's computation in the forward direction is a combination of function applications and matrix multiplications. If the input to the network is a vector $\vec{x}$, the activation function in layer l is $f^l(\ )$, and $g(\ )$ represents the total transformation from input to the output, we can write

$$\mathbf{g}\left(\vec{x}\right) = \mathbf{f}^{L-1}\left(\mathbf{W}^{L-1}\left(\cdots\left(\mathbf{W}^2\left(\mathbf{f}^1\left(\mathbf{W}^1\vec{x}\right)\right)\right)\right)\right) \tag{4.54}$$

Thus, the entire neural network can be thought of as a sequence of vector-matrix or tensor computations. GPU machines do not have to compute in terms of layers and nodes, just in terms of tensors and function applications.

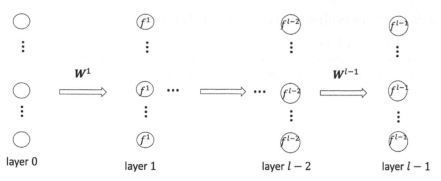

FIGURE 4.33: Feedforward neural network with L layers

Visualization in terms of layers and nodes is to help humans understand and think about neural networks.

A training set is a set of examples $\mathbf{D} = \left\{ \left\langle \vec{x}^{(i)}, y^{(i)} \right\rangle \right\}, i = 1 \cdots N$. For an input-output pair $\left\langle \vec{x}^{(i)}, y^{(i)} \right\rangle$, the loss (error) of the model is a function of the difference between predicted output $g(\vec{x}^{(i)}) = \hat{y}^{(i)}$ and the target output $y^{(i)}$, written as

$$l\left(y^{(i)}, \mathbf{f}^{L-1}\left(\mathbf{W}^{L-1}\left(\cdots \left(\mathbf{W}^2\left(\mathbf{f}^1\left(\mathbf{W}^1 \vec{x}^{(i)} \right) \right) \right) \right) \right) \right). \quad (4.55)$$

In order to compute the loss, one starts with the input $\vec{x}^{(i)}$ and computes $\mathbf{W}^1\vec{x}^{(i)}$, etc., moving forward in the network till the last layer. As discussed earlier, this loss function is computed for each input-output pair individually and all weights are updated for the learning algorithm called Stochastic Gradient Descent (SGD). For Mini-batch Gradient Descent, the loss is computed for each individual input-output pair in a (mini-)batch; the individual losses for the batch are added and averaged; and all the weights are updated using backpropagation considering the average loss for a batch. An epoch of learning consists of learning from all batches (or, all examples) in a training dataset.

We have not discussed how backpropagation can be expressed as matrix-vector or tensor computations. It is left as an exercise.

4.9   Conclusions

Artificial neural networks and the deep learning version of ANNs have recently been able to solve a variety of large-scale machine learning problems, causing great excitement in the academic research community as well as developers and practitioners of machine learning in large corporations such as Google, Facebook, Amazon, Microsoft and others. Small companies and technology start-ups of all kinds are also leveraging machine learning in their products.

This chapter has introduced the basic concepts in artificial neural networks and deep learning. On the topic of deep learning, we have discussed CNNs only. There is another common variety of deep learning for sequences such as words in a sentence or a document. We have not covered this variety of deep learners called Recurrent Neural Networks (RNNs). In addition, we have not covered recent developments such as attention mechanisms, residual connections, transfer learning and multi-task learning in neural networks. The advanced learner can read recent books on deep learning. In addition, there are a large number of well-prepared web pages on these topics. Further, there are many videos on YouTube by well-known professors, researchers and practitioners that are likely to be of immense help to any student of deep learning.

Exercises

1. An artificial neural network may have hundreds of thousands, millions or even billions of parameters or weights. Neural networks are trained on a dataset to obtain optimal weights so that the neural network transforms an arbitrary input to an output to perform prediction. There are many issues in making a neural network work! One such issue is initialization of weights. The general wisdom is that network weights are initialized randomly. However, random initialization may not always work if we want to obtain to the network to learn well. What are issues in initialization of a neural network? Describe a technique that has been proposed to initialize neural networks.

2. For backpropagation to work, it is desired that an activation function is differentiable. Two differentiable activation functions that are commonly used are tanh and softmax. Obtain the derivatives of these two functions to show that they can be conveniently used for backpropagation.

3. For backpropagation to work, we also need a loss function that is differentiable. We have discussed the squared error loss in this chapter. Another commonly used loss function is cross-entropy loss. What is the cross-entropy loss function? There are two versions of it: binary cross-entropy loss and multi-class cross-entropy loss. Give a mathematical formula and show that both versions of it are differentiable.

4. The ReLU activation function is commonly used in Convolutional Neural Networks. However, it is not differentiable, causing an issue. Why is

ReLU still so commonly used? How can it be effectively differentiated so that it can be used in backpropagation?

5. Gradients are important to describe how neural networks work. Efficient and effective computation of gradients is important for implementation of neural networks. However, since neural network weights are initialized randomly and inputs may vary in magnitude, the gradients could become very high resulting in what is called the *exploding gradient problem,* and sometimes the magnitudes can become very small resulting in the *vanishing gradient problem.* Discuss how the exploding and vanishing gradient problems can be handled in a practical neural network implementation.

6. Gradient descent is essential for neural networks to function. Each data point can be taken individually to compute the gradient, or the entire dataset can be used to compute an average gradient over all points. The first approach is called *stochastic gradient descent,* and the second is called *batch gradient descent.* As a compromise, a small number of data points, say 128 or 256, can be taken in small or mini batches. Compare these three approaches to gradient descent in terms of efficiency and convergence.

7. The main essence of machine learning is to learn an optimal values for a set of parameters from a training dataset such that the learned model is generalized. In other words, the learned model does not simply memorize the data examples it has seen, but can do well on predicting results for unseen data. Generalization sometimes needs help in the form of what is called *regularization.* Describe ways to regularize the optimization of weight (parameters) in an artificial neural network.

8. Dense feed-forward neural networks as well as convolutional neural networks take a vector of input (or a 2- or 3-dimensional matrix for images) and produce an output which is usually a vector. However, often we have to learn how to transform i) a sequence of input to a single output, or ii) a single input to a sequence of outputs, or iii) a sequence of input to a sequence of output. How can neural networks be architectured so that learning is possible in such situations? Describe an architecture that can process a sequence of inputs or outputs. How can such an architecture be trained?

Chapter 5

Reinforcement Learning

In Chapter 1 of this book, we discussed regression whereby given a dataset with numerical predictor variables and a numerical dependent variable, a program learns to predict the value of the dependent variable for unseen values of the independent or predictor variables. This chapter also introduced ideas of optimization and regularization. In Chapter 2, we presented the idea of tree-based machine learning. Trees can be learned from a dataset with either all numeric predictor variables, or all non-numeric or discrete predictor variables, or a mix of numeric and non-numeric independent variables. The dependent variable can be numeric or non-numeric as well. If the predicted variable is non-numeric, taking a small number of values, it can be considered a label and the process of machine learning is called classification. Each label is considered a class. Tree-based machine learning algorithms such as boosted trees are usually the best performers for featured datasets, where a data example is described in terms of a small number of independent variables called features. In Chapter 3 of the book, we discussed artificial neural networks, including deep learning. Deep learning has excelled in solving regression and classification problems for "raw" datasets such as images as well as text. In such a dataset, "features" or characteristics of the data are not extracted a-priori. Raw values for a data example, such as all pixels of an image become the descriptors of the example, whereas in a featured dataset, features such as edges and shapes may have been extracted before feeding the data to a machine learning algorithm. Deep learning automatically extracts features at various levels of abstraction. Deep learning can be used for regression as well as classification.

Both regression and classification are examples of supervised learning. In supervised learning, each example in the dataset has an associated true label. The label is said to be provided by a "supervisor", often a human in classification problems. For regression, it is likely that the label values, which are numeric, come from measurement with a tool or a sensor. For example, the label can be a measured temperature, pressure or an angle.

In this chapter, we introduce reinforcement learning, which is different from both supervised and unsupervised learning. Here, the assumption is that there is a software-based agent situated and navigating in a space of states, which can perform a certain action out of a set of enumerated actions at each state, in its quest to get to a goal state from a (start) state, where the last transition gives it a high reward. For example, in learning how to travel

DOI: 10.1201/9781003002611-5

successfully through a maze, a software-guided agent such as a robot, can take an action such as moving left, right, forward or backward at various time points and locations; it gets a high reward only when it gets to the goal state, i.e., outside the maze, not even a moment earlier, just like a racer does not win even if it has a fraction of a centimeter left to get to the finish line. As another example, in going from point A to point B, a software-controlled autonomous vehicle may need to decide between stopping or making a left or a right turn, at various decision points as it moves. It gets a high reward only when it transitions or gets to point B; it does not get a reward if it misses the goal point B even by a small margin. Another way to think about is a strict teacher who gives a student a score (reward) of 100 only if the answer to a test question is completely correct, and nothing at all even if the answer is 99% done, and thus, no partial credit or reward of any kind except at the very end.

Another difference between reinforcement learning and other kinds of learning is that reinforcement learning necessarily involves an agent learning to perform a sequence of actions toward a goal and achieve the goal. In this setup, it is neither appropriate nor useful to think of an agent performing one action, but a sequence—some of which may be "good" and some "bad", but altogether the sequence produces a high reward at the end. In classification, which is a form of supervised learning, each training or testing example is labeled or classified on its own, without any concept of a sequence.

5.1   Examples of Reinforcement Learning

To introduce the idea of reinforcement learning, let us start with a few simple examples, learning to navigate through a maze, playing a simple 1-person game called 8-puzzle, and playing Atari games.

5.1.1   Learning to Navigate a Maze

Consider a simple maze as given in Figure 5.1. We want to build a software agent that starts at a given location in the maze and learns to obtain a path to the goal. In this example, there is only one start state, although in some cases, several start states are natural. The shaded cells are obstacles. If we number the cells row by row from left to right, we can think of the agent being in one of the states in a set of states S,

$$S = \{s_1, s_2, s_3, \cdots, s_{16}\},$$

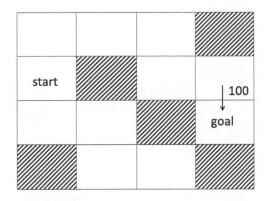

FIGURE 5.1: A simple 4x4 maze

where each cell is considered a state. When at a certain state, the agent can perform one of a set of actions A,

$$A = \{\leftarrow, \rightarrow, \uparrow, \downarrow\},$$

where the arrows represent the direction of motion of the agent. Note that all actions may not be possible in all states since the agent could fall off of the board. The goal of reinforcement learning, in this case, is to learn to find a path or a sequence of state transitions from start state s_5 to goal state s_{12}. An ancillary goal may be for the agent to learn the best action to perform when it finds itself in any cell of the maze. We assume that the agent has a model of the environment, i.e., it knows the actions it can perform in various states with associated probabilities for the actions. Or, we can also assume that the agent is model-free, and it starts with a blank slate without any knowledge of probabilities of transitions at various states. In both cases, the agent performs actions on its own to learn the best action to perform in each state so that it can achieve the objective of landing at the goal state s_{12}.

We see that reinforcement learning is about learning to perform a sequence of actions to achieve a goal. The agent is in a certain environment and performs actions in a sequence as long as necessary till it arrives at the goal, when it receives an immediate reward. In general, there could be multiple states where the agent can receive rewards, which can be of various magnitudes. In addition, rewards can be negative as well, i.e., actually punishments. In this example, there is only one positive reward, which the agent receives when it arrives at the designated goal state. In this example, the agent gets (an immediate) reward of $+100$, say, when it arrives at the goal state. At every other state, it gets (an immediate) reward of 0 on arrival. It is possible that in other scenarios, there are positive and negative rewards of various magnitudes.

Based on one positive reward, which it gets on arrival at the goal state, the agent needs to learn what actions it should perform at the states along the way so that it can arrive quickly at the goal state. It may learn by moving randomly

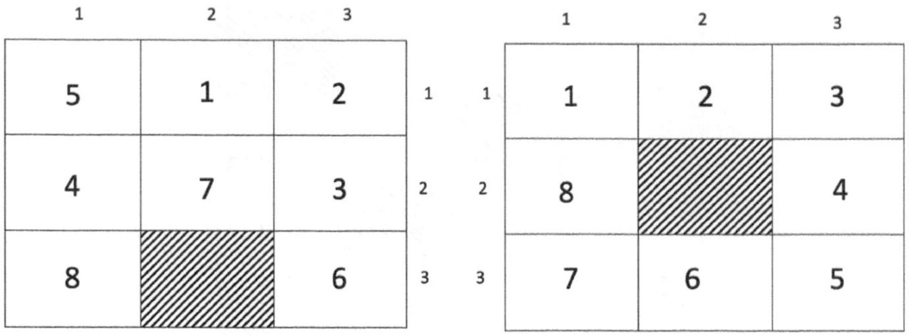

FIGURE 5.2: 8-puzzle: a) A random initial state, and b) the final desired state

or using some other strategy, from start state to goal state, many many times. Initially, its movements may be uninformed, and (almost) "blind", but as it tries more and more times to go from start to goal, it learns values of states and actions better; after some time, it becomes proficient in doing so.

5.1.2   The 8-puzzle Game

The second example is a game called 8-puzzle, where there is a square board with equal sized pieces that can be moved by sliding. There is one spot that is empty on the board. On each piece, there is an inscribed integer from 1 to 8, each number occurring only once. One can have variations of the 8-puzzle game such as ones with 15 pieces or 24 pieces. In general, the board size is $n \times n$ where n is a small positive integer. The pieces are jumbled up initially, and the game involves moving the pieces around so that finally they are in a specific sorted order. Figure 5.2 shows a random initial configuration of an 8-puzzle board and the final finished board.

To write a program that learns to play this game, it is helpful and simplifying to think of the shaded empty spot as the only moving part, instead of the pieces with numbers written on them. This decreases the moving "parts" to just one, instead of eight. The actual pieces move when we "move" the empty spot. This empty spot can, in general, move in 4 different ways. If we think of a software agent playing the game, the agent can perform four different actions: move the empty spot left, right, up or down. Like before, we have a set of actions,

$$A = \{\leftarrow, \rightarrow, \uparrow, \downarrow\}.$$

Note that, like before, certain actions may not be performed in certain states as the empty spot or the actual pieces must not fall off the board. In this case, we can think of each configuration as a state of the game; thus, the state representation is more complex. Therefore, a state of the game is a 2-D matrix of integers with an empty cell in the matrix. Figure 5.2 shows two

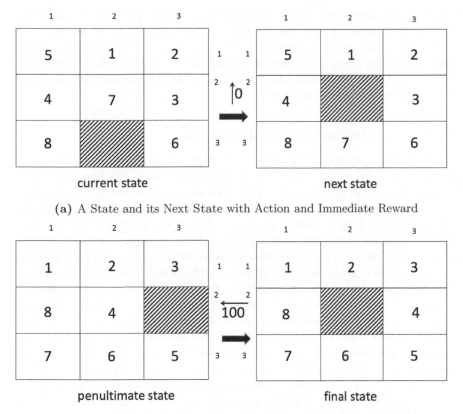

(a) A State and its Next State with Action and Immediate Reward

(b) A Penultimate State and the Final State

FIGURE 5.3: States, Next States, Actions and Immediate Rewards

states, and Figure 5.3 shows two pairs of additional states, where one leads to the other, with actions and immediate rewards shown. The action is indicated on top of the fat arrow, and the immediate reward below it. The first pair is a random state and its successor state, and the second pair shows a penultimate state and the goal state. In the first pair, the agent moves the empty spot up (action performed is ↑) and receives immediate reward of 0. In the second pair, a movement of the empty spot to the left leads to the final state where the agent gets an immediate reward of +100, say. Transition to only the final state gets a positive reward. Transitions to any other state using any other action results in an immediate reward of 0.

Once again, the agent learns to play by actually playing. It can mess up the pieces to create a random initial state and then play till it gets to the final finishing configuration. Assuming the agent perseveres, initially it may take a long amount of time playing in an uninformed, dumb manner. But, as it plays more and more times, from beginning to end, it becomes "smarter" and learns to appreciate good states and good actions in context, and finally after arduous work, it becomes a good or even an excellent player that can make moves more or less flawlessly in every step. In lieu of actually playing the game physically, being a software agent, the agent can play a simulated game, where everything happens quickly in software, without the outward display of a board with numbers that move.

5.1.3   Learning to Play Atari Games

Atari, Inc., pioneered the production and popularization of video game consoles in the 1970s and 1980s. The Atari 2600 was one of the first video game consoles that could be adapted to play various games by swapping cartridges. The console could be bought with various controls such as joysticks, switches and paddles to play a variety of games. Examples of games on the Atari 2600 were Beam Rider, Breakout, Enduro, Pong, Q*bert, Sequest and Space Invader. The player(s) would see the current state of the game on the computer screen, and then manipulate the objects in the scene with the controllers to play. For example, in single-player Pong, the player would move two paddles on the two sides of the console to play ping pong. In the simplest variation of Pong, the paddles could move only up and down on the margins of the board.

Researchers from DeepMind Technologies have developed programs that learned to play seven Atari games, all in software simulation, well using reinforcement learning. In 2013, they were able to outperform all competing programs in six of seven games mentioned above, and surpass human performance on three. In this case, the program was given console images of size 210×160 pixels with 8-bit or 128 colors. The images were clipped and downsampled by converting to 84×84 gray pixels. The program played on its own and learned to play well playing a very large number of examples alone. The actions to be performed included moving the two paddles up and down (they are constrained to move vertically only) by a certain distance. The number of states are potentially very large since a state is represented by the image of a 84×84 pixel frame of the video. The action to be performed needs to specified in terms of a continuous positive or negative distance moved vertically within a given range. The immediate rewards come when a player scores a point, and a bigger reward comes when a player wins the game. The reward values can be set up arbitrarily, say 1 for each won point, -1 for each lost point, and 10 for a game won and -10 for a game lost. Beside these, for all other actions, the immediate rewards are 0. The assumption here is that one player is an AI reinforcement-based leaner, and the other is a human player. However, as mentioned earlier, there were many variations to the Pong games that Atari provided in its cartridges. Between 1 and 4 players could play the variations of the Pong games.

Again, the software player would start from a certain state of the game and play till it wins or loses, receiving positive or negative rewards. It would, in some sense, contemplate and think what actions work well when and what

(a) Atari Console with Joystick and Four Switches from 1980-82 (from Wikipedia)

(b) Atari Console Showing Two Ping-pong Paddles, the Ball and the Scores for Two Players (from Wikipedia0

FIGURE 5.4: Atari Console and Pong Board

do not, once it has finished a game. The next time around, it would again play from beginning to end, but potentially would play a little better, by becoming better at its choice of actions at different states of the game. After millions or billions of games, it would become a really good player, even a world champion. Since everything happens in software, individual games can be played very quickly, but it would still take days or weeks of playing on high-powered computer systems, to go from being a dumb novice to a player that plays at the level of a world champion, shattering records.

5.2 The Reinforcement Learning Process

In reinforcement learning, the agent resides, performs actions and receives immediate rewards in an environment. This is shown in Figure 5.5. The agent learns in episodes. An episode involves the agent starting at a certain (permitted) state, performing an action, obtaining an immediate reward if any, and repeating the process till a goal (high-reward, positive or negative) state that signals termination, is reached. Note that there may be several start states, or the start state may be even random, depending on the problem to be solved. The sequencing of states, actions and rewards is shown in the following page.

$$s_o^{(i)} \xrightarrow[r_0^{(i)}]{a_0^{(i)}} s_1^{(i)} \xrightarrow[r_1^{(i)}]{a_1^{(i)}} \cdots\cdots s_{n_i-1}^{(i)} \xrightarrow[r_{n_i-1}^{(i)}]{a_{n_i-1}^{(i)}} s_{n_i}^{(i)} \qquad (5.1)$$

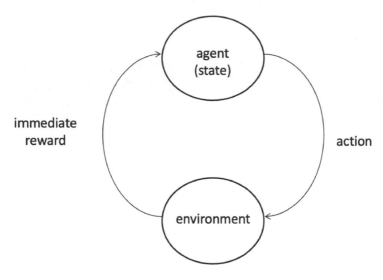

FIGURE 5.5: The Reinforcement Learning Environment

The same sequence can be written in the following way also.

$$s_o^{(i)}, \ a_0^{(i)}, \ r_0^{(i)}, \ s_1^{(i)} \ \cdots\cdots \ s_{n_i-1}^{(i)}, a_{n_i-1}^{(i)}, \ r_{n_i-1}^{(i)}, \ s_{n_i}^{(i)} \tag{5.2}$$

The agent starts with a state $s_o^{(i)}$, the first state in episode i. It performs an action and gets an immediate reward and transitions to a new state. The top of the arrow shows an action performed, and the bottom shows the immediate reward obtained. The superscript (some subscripts as well) i represents the ith episode of training. It can be dropped to make the notation simpler and write the sequence as given below.

$$s_o, \ a_0, \ r_0, \ s_1 \ \cdots\cdots \ s_{n-1}, a_{n-1}, \ r_{n-1}, \ s_n \tag{5.3}$$

Here, s_n is the goal state in this episode. The agent gets an immediate reward r_{n-1} before reaching the goal state.

Assume in episode i of training, the agent starts at a certain state and performs an action. The environment itself or a "teacher" (or a coach or observer) in the environment gives the agent an immediate reward. In an episodic learner, the process of performing an action, receiving an immediate reward and transitioning to a new state continues for a while till a goal state is reached. The immediate rewards can be given by the environment itself or by a coach or referee, who can be thought of as supervisor. Thus, reinforcement learning can thought of as supervised learning in a way. However, the rewards are rare, and as a result, most actions have an immediate reward of zero, which can essentially be considered neither supervised nor unsupervised. Therefore, reinforcement learning has the characteristics of both supervised and unsupervised learning. The agent keeps note of its state. The agent's representation of state can be simple or complex. In addition, the agent's representation of what a state is, may be limited in the sense that it may be a limited version of what the real state is, because the agent is not omnipotent. In other words, the agent does not see every characteristics of the environment. The real state in the context of the environment may be much more complex than what the agent is able to record, keep and (needs to) consider. When we talk about states, it is the agent's version of states.

In reinforcement learning, the learner wants to learn what is called a *policy*. A policy is a function that tells the agent what action(s) to perform in which state so that the agent can maximize collected or cumulative rewards along the way to the goal. We can write

$$\pi : S \to A^+ \tag{5.4}$$

where π is the policy function, S is the set of states and $A^+ = A \times A \times \cdots A$ representing one or more actions. That is, a state can be mapped to more than one action, i.e., a subset of actions available in set A. In other words, it is possible that several actions are equally permissible and worthy in some states and one can be chosen randomly. However, we can simplify our thinking of reinforcement learning by assuming that it is a mapping

$$\pi : S \to A \tag{5.5}$$

where the agent learns to perform one action at each state. If several actions are equally optimally permissible or valuable, i.e., best, the agent can choose one of the alternative best actions randomly, without prejudice.

Although the ultimate goal is to learn a policy, there are different ways to do so. All of these techniques involve learning so-called *values*. One value-learning approach may learn values for every state in the state space first. It uses the state values to derive the policy function. Another approach may learn values for each state-action pair $< s, a >$, i.e., how valuable a certain action a is in a certain state s. It obtains the values of the states from the values of $\langle s, a \rangle$ pairs. The value of a state s represents how valuable the state is, considering that the objective of the agent is to get to a goal state, collecting immediate rewards, if any, along the way. In other words, the valuation of a state quantifies in some manner how prepared to get to or close to a high-reward achieving goal state the agent is, being in this state s. A state s that is close to a goal state, by going to which the agent gets a high reward, is more valuable than one far from such a state. Similarly, the value of an action a in state s represents how valuable the state-action pair is, considering that the agent wants to get to a goal state, where the agent gets a high reward in the final transition. For example, being in a high-valued state and performing a high-valued action in such a state is more desirable than being in a high-valued state, but performing an action that may take the agent away from a goal state. The first approach may be called *state-value learning* and the second approach may be called *state-action value learning*.

If there are multiple reward-obtaining state transitions on the path to the goal, the agent wants to maximize the cumulative reward along this path. The cumulative reward is the summation, in some (weighted) manner, of all immediate rewards picked up along the way. For an example, that does not quite fit our discussion here, but still is quite telling, we can think of a game of soccer, where although the ultimate goal is to win the game when the final whistle blows (i.e., receive a large immediate reward at the very end of the game), a team obtains smaller positive rewards when it scores goals against its opposing team, as well as smaller negative rewards when goals are scored against it. These rewards may matter if the game ends in a draw. A soccer team is comprised of several individual agents playing collaboratively against (i.e., competitively with) another collaborative team, and hence it represents a much more complex agent-environment interaction scenario. However, if each of the two teams had only one player, the situation becomes much more comparable to the examples discussed in this section.

On another note, in discussing these approaches, the initial assumption is that the number of states and the number of actions are both small. The approaches can be potentially expanded to really large, even continuous spaces. We will discuss such extensions briefly later in the chapter.

5.3   Rewards—Immediate and Cumulative

The reinforcement learning process is driven by the concept of rewards. When an agent performs an action a_i at a certain state s_i, it gets an immediate reward r_i, as indicated above. Immediate rewards can be added to compute a cumulative reward.

5.3.1   Immediate Reward

To make things very simple, let us assume that a soccer playing agent is learning how to score a goal. Thus, an episode of training consists of an agent starting from a random cell (say, on a grid) in the soccer field, performing various actions till it comes to a state (penultimate state s_{n-1}, say) where it stands a few feet in front of the goal post and kicks (action a_{n-1}, say) the ball to score a goal (transition to a goal state, s_n, say). The action of kicking the ball receives an immediate reward of r_{n-1}, say, of value $+100$. In soccer, states and actions are not straightforward to define. If the soccer field is divided into a rectangular grid, say of size 120x100, each of the 12,000 cells can potentially be a state. The actions can be the agent moving with the ball from one cell to another in different directions, and kicking the ball with different amounts of force in different directions. These actions become difficult to define and may have to be parameterized, with parameters such as angles and amounts of force. Let us not dwell on such complexities at this moment.

As a simpler example, we can consider the maze in Figure 5.1. The goal state is indicated to be in the third row and fourth column in the 4×4 grid. The cell $(3, 4)$ in row 3, column 4 is the final or penultimate state s_n, say, in this case. The state $(2, 4)$ can be thought of as the penultimate state and let us call it s_{n-1}. When in this state s_{n-1}, if the agent goes down by performing the action $\downarrow$, the agent gets an immediate reward r_{n-1}, say of $+100$ and gets out of the maze, by going to the goal state s_n. In the maze example, the cell $(2, 3)$ is the state s_{n-2}, the agent must get to before going right (action $\rightarrow$ or a_{n-2}) to the penultimate state s_{n-1}. In the way we discuss reinforcement learning, we assume that there is no immediate reward obtained by action a_{n-2} in state s_{n-2}. That is $r_{n-2} = 0$. In the maze example, all immediate rewards are zero, except on the $\downarrow$ to cell $(3, 4)$. In the soccer example, where the agent is not playing a full game, but learning how to score a goal on the opponents, only the last action receives a positive immediate reward; all other actions at all other states get immediate reward of 0.

Not all immediate rewards, for all actions except for the goal-transiting action at a pre-goal state, have to be 0. For example, if the maze had a cell, which was a deep well, an action at any neighboring cell, transitioning to the well, could have been given a high negative reward, say -100. For a soccer playing agent, an action of kicking the ball through the goal posts of its own side could net a negative reward of say, -100. The soccer playing agent could possibly get a small negative reward, of say -1 for kicking the ball outside the boundaries of the field for at action that leads to an 'out of bound' ball.

5.3.2 Cumulative Reward

An episodic agent, starting at a start state, may perform a (long) sequence of actions before getting to a goal state. Assuming there are, at least some non-zero rewards along the way, it may pick up several immediate rewards before reaching a goal state s_n, where it receives a large reward. When we talk about cumulative reward, it refers to the sum of all immediate rewards the agent, starting at state s, gets along the way, culminating at the goal state. Let us denote the cumulative reward an agent collects when it starts at state s by R_s. For a sequence as in Equation 5.1, 5.2, or 5.3, we can write

$$R_0 = \sum_{i=0}^{n-1} r_i, \tag{5.6}$$

where we directly sum all the immediate rewards. We can also write R_1, cumulative reward starting at state s_1 as

$$R_1 = \sum_{i=1}^{n-1} r_i, \tag{5.7}$$

and so on and so forth.

It is also possible, that in some situations, we may also consider immediate rewards obtained early in the episode to be more valuable than later in the episode. We will discuss when such a computation is appropriate in a later section. In such a set-up, we need to introduce the concept of discounting future or later rewards. Discounting or lowering the value of later rewards is done using a so-called *discount factor* γ such that $0 \leq \gamma \leq 1$.

The cumulative reward for the suffix (state-action-reward transitions starting from a later state to a goal state) of an episode, starting at state s_t in such a scenario is given as

$$R_t = r_t + \gamma r_{t+1} + \gamma^2 r_{t+2} + \cdots + \gamma^{T-t-1} r_{T-1} \qquad (5.8)$$

$$= \sum_{i=0}^{T-t-1} \gamma^i r_{t+i} \qquad (5.9)$$

assuming the episode ends at time T. Here, the state s_t is t steps from the beginning of the episode. For example, the cumulative reward starting from state s_0, assuming that we start at $t = 0$ is

$$R_t = r_0 + \gamma r_1 + \gamma^2 r_2 + \cdots + \gamma^{T-1} r_{T-1}. \qquad (5.10)$$

5.3.3   Discounting Future Rewards

The question arises, why are future rewards discounted in computing cumulative reward, when starting from a state s? The discount factor γ is such that $0 \leq \gamma \leq 1$.

The term *discount factor* is commonly used in economics or mathematics, where it is used to compute the present value of future happiness, incomes or losses. For example, if a person wants to get a loan from a bank now, and he/she does not have any income now but anticipates a large income, say of a million dollars exactly one year from now, the bank may be willing to give him/her a loan not of 1 million dollars, but a little less. It is possible that the bank uses something like a discount factor γ to compute the loan it is willing to give at present. If $\gamma = 0.9$, considering a period of 1 year as a unit, the bank may be willing to give a loan of $\$1M \times 0.9 = \$900,000$ now. If the income of $\$1$ million is anticipated exactly after two years, the bank may be willing to give a loan of $\gamma^2 \times 1,000,000 = (0.9)^2 \times 1,000,000 = \$810,000$ loan now.

If $\gamma = 0$, only the immediate reward, when starting from a state s and performing an action a is considered. Such an agent is considered myopic or short-sighted since it ignores any rewards it may obtain in the future completely. In terms of the bank loan example, it essentially says that to get a $\$1$ million loan now, a person seeking the loan must have an income of 1 million dollars now. Any anticipated incomes in the future are not considered by the bank.

On the other hand, if $\gamma = 1$, and the person seeking a loan has yearly anticipated incomes of 1 million dollars at the end of each of years 2, 3, 4 and 5, the bank may be willing to loan \$5 million now, by adding all anticipated incomes (rewards) within a horizon of five years. However, money usually depreciates over time. If the bank assumes a discount (depreciation rate) of $\gamma = 0.9$, the bank may be willing to loan now a total amount of $\$1M + 0.9 \times \$1M + (0.9)^2 \times \$1M + (0.9)^3 \times \$1M + (0.9)^4 \times \$1M = \$1 + 0.9 + 0.81 + 0.729 + 0.6561 = \$4.0951M$, not \$5M. This is because incomes are anticipated, not yet real; although they are likely, things could change in the economy and the environment. For example, there would be a recession, a bad disease may afflict the person, or there may be a war that affects the person's income abilities.

The same idea is used in reinforcement learning. When computing cumulative reward at a state s, future immediate rewards are not added at full value, but with a discount. Thus, although during training in actually completed episodes, future (later) rewards are actually obtained since the episode has been performed. However, when an agent that has gone through some episodes of training, performs actions based on this knowledge, there is some level of uncertainty associated with future rewards because the agent operates in an environment and there may be unanticipated changes in the environment, making certain actions performed in certain states during training, impossible in actual functioning. For example, if we have an episode laid out from beginning to end for a maze-traversing agent, the contribution of the immediate reward at the penultimate state to the cumulative reward at an earlier state would be lower than at a later state. This essentially says that earlier in an episode, the chance of going astray is higher (or there are many alternative or possible pathways at an earlier state) that when near the goal.

5.4 Evaluating States

The power of reinforcement learning comes from a trained agent being able to anticipate the cumulative reward it would get starting at a certain state s. The purpose of the agent during training is to learn to predict what cumulative reward it may get, starting at a certain state s; better if it can do so at the very onset of an episode. To be able to perform such predictions later, during training the agent starts many episodes at state s, collects rewards, computes cumulative reward for each episode, and averages the cumulative rewards. When we compute cumulative reward for an episode during training, we use the discount factor as discussed above. This is because although the agent learns from actually "performed" episodes, ultimately a trained agent will have to perform actions, when it finds itself in various states in the state space. The agent's knowledge is "always" incomplete in the sense that at any time, it has learned from only a limited number of episodes, and the number of training episodes potentially could continuously increase over time. The idea of discount factor in reinforcement learning is similar. However, the expectation is that after ra certain large number of episodes, the learned values converge.

In reinforcement learning, the agent's ultimate objective is to learn what actions are best to perform in what states. In turn, this enables the agent to perform a sequence of actions in pursuit of a goal , expressed as acquiring high cumulative reward. In can learn so indirectly, first learning how to evaluate states, or state-action pairs. We will discuss how to evaluate states here.

To learn to evaluate a state s, as mentioned earlier, the agent can perform a number of episodes of training starting at state s, compute R_s for each episode and average theme. If $V(s_t)$ is the value of a state s_t, we write $V(s_t) = \mathbf{E}(R_{s_t})$ where $R_{s_t} = \sum_{i=1}^{T} \gamma^i r_{t+i}$. Here $\mathbf{E}$ stand for expected value; in practice it means "on average". The value of a state quantifies the prospects of a state to get cumulative rewards if an agents starts from the state. The value of state s_t can be written as

$$V(s_t) = \mathbf{E}(R_{s_t})$$

$$= \frac{1}{J} \sum_{j=1}^{J} \sum_{i=0}^{T_j} \gamma^i r_{t+i}. \qquad (5.11)$$

Here, T is the length of an episode, also called the horizon. We can keep it constant if we like. In games and problems we have been discussing, the value of T is likely to vary from episode to episode, depending on the path taken. That is why we use T_j as the number of time steps of the horizon length in episode j. Assuming we perform J episodes of training starting at state s_t, we compute the cumulative reward actually obtained at the state for each of the episodes, add them and obtain the average. We can clearly see that the estimate of $V(s_t)$ improves as the agent performs more and more episodes. After the first episode of training starting at state s_t, it is likely to have a crude estimate of $V(s_t)$. But, once the number of episodes goes up to $2, 3, \cdots$, the estimate of $V(s_t)$ is likely to improve and eventually converge, under certain assumptions.

5.4.1   Estimating $V(s)$ Given Policy π

Let us assume that a policy function π is given to an agent. In other words, the agent knows what action to perform in what state. The state space is "small" and enumerable. An algorithm to find the value of a state in the state-action-reward state space may consist of the agent starting from a (random) start state and going through an episode of either simulated or actual experience till it reaches a goal state, following policy π. Thus, the agent performs actions in states and gets rewards, and continues till a goal state is reached. It keeps a count of the number of times it starts from a certain state. For each new episode, it computes the cumulative reward starting from that state, and obtains the average cumulative reward starting from that state. The average value is the current value of the state. The estimate gets better and better as more episodes are started from a specific state. Algorithm 5.1 is the pseudocode for such a process.

Algorithm 5.1: Estimate Values of States Under Policy π

Input: A policy π to be evaluated
Output: Values of $V(s)$, $\forall s$
```
/* Initialize                                          */
```
1 **for** *each state $s \in S$* **do**
2   Set $V(s)$ to a random value;
3   Set $C(s)$ to 0 ; /* Count of episodes starting at state s */

4 **Loop forever**
5     $s_0 \leftarrow$ a start state for the episode; Pick randomly if there are several start states;
6   Generate an episode following π for T time steps, when a goal state is reached:
7        $s_0, a_o, r_o, s_1, a_1, r_1, \cdots s_{T-1}, a_{T-1}, r_{T-1}, s_T$;
8     $R(s_o) \leftarrow \sum_{k=0}^{T-1} \gamma^k r_k$;
9     $V(s_o) \leftarrow \frac{V(s_o) \times C(s_o) + R(s_o)}{C(s_o) + 1}$;
10    $C(s_o) \leftarrow C(s_o) + 1$

This algorithm is very simplistic. It keeps a vector $V(s)$, $s = 1 \cdots N$, where V is the valuation of a state and N is the number of states in the state-action-reward space. It keeps another vector $C(s)$, $s = 1 \cdots N$, where it keeps the count of the number of episodes starting from each state s in the space.

The algorithm loops forever. In practice, it can loop for a certain (large) number of times, and learn valuations of states, and then use the valuations obtained experimentally as the estimate of actual valuations of states. It can intersperse actually performing actions with updating valuations of states. This algorithm is simple, and therefore, beset with several problems. Let us enumerate some of the issues with this algorithm and contemplate ways to improve it.

An obvious problem with Algorithm 5.1 arises because it assumes that the policy π is given. This means that when the agent starts from a start state, the agent is always going to follow the same sequence of states and obtain the same rewards. This assumes that the transitions and rewards, i.e., the dynamics of the system, are deterministic. In a deterministic system, when an agent performs an action at a state s, it always transits to the same next state s', and when an agent performs an action a in state s, it always gets the same reward r. Thus, this algorithm, as it is written, learns to obtain valuations of the start states and no other states in the deterministic case. As a result, this algorithm does not really do what we set out to do, which is to obtain valuations of all states in the space in the deterministic case, whereas it may work much better in a non-deterministic situation.

To rectify the situation, in each episode of training, we can update $V(s)$ value for not just the start state, but for each state s that occurs along the way from a start state to a goal state. If there is only one start state and the policy π recommends only one action per state, one episode of training will obtain the values for all states along the path from start to goal. However, it is possible that a more general policy may provide several actions in certain states because the actions are equally likely. In such a case, depending on which action is chosen (possibly randomly) in a state s, the suffixes (or later sequence of states) of the path beyond s may come out different along different paths, and averaging cumulative rewards along different paths will update the valuations of s to a more realistic value.

Algorithm 5.2: Estimate Values of States under Policy π

Input: A policy π to be evaluated
Output: Values of $V(s)$, $\forall s$
```
/* Initialize                                              */
```
1 **for** *each state $s \in S$* **do**
2   Set $V(s)$ to a random value;
3   Set $C(s)$ to 0 ; /* Count of episodes starting at state s */

4 **Loop forever**
5     $s_0 \leftarrow$ a start state for the episode; Pick randomly if there are several start states;
6   Generate an episode following π for T time steps, when a goal state is reached:
7        $\langle s_0, a_o, r_o, s_1, a_1, r_1, \cdots s_{T-1}, a_{T-1}, r_{T-1}, s_T \rangle$;
8   **foreach** *state $s \in \langle s_0, a_o, r_o, s_1, a_1, r_1, \cdots s_{T-1}, a_{T-1}, r_{T-1}, s_T \rangle$* **do**
9   Let k be the index of state s;
10        $R(s) \leftarrow \sum_{j=1}^{T-k-1} \gamma^j \, r_{k+j}$;
11        $V(s) \leftarrow \frac{V(s) \times C(s) + R(s)}{C(s)+1}$;
12        $C(s) \leftarrow C(s) + 1$

In this modified algorithm, the agent updates the value of each state along the path of the episode from start to goal. This means that in each episode up to T states may have their valuations updated, making Algorithm 5.2 a lot more efficient than Algorithm 5.1. However, we can point out some issues with this algorithm to improve it further. First, it performs the cumulative reward computation in a forward direction, traversing the states from the direction of the start state to the goal state. After completion of an episode, if instead the cumulative rewards are computed in the backward direction, from the penultimate state—the state immediately prior to the goal, the computation can be written much more efficiently, dispensing with most of the summations.

Another problem that may arise is that in the list of sates in an episode, there may actually be loops. If a certain state occurs more than once, the agent may never reach a goal state, being in an infinite loop, especially if there is only one action per state in the policy. If a policy allows more than one action in a state, and one of these actions is chosen randomly, the episode will end, although it may take a while. A general solution to this problem may be that the agent does not always wait till an episode arrives at a goal state, but carries out the episode to a certain horizon H, which is a fixed positive integer. For example, if $H = 100$, it carries out the episode at most 100 steps; if it does not end at a goal state by this time, it abandons calling it an unfinished episode, but it computes the cumulative reward and performs the averaging anyway. If there are one or more non-zero rewards along the way before the episode hits the horizon, there may be a non-zero cumulative reward. The modified algorithm is given as Algorithm 5.3.

Algorithm 5.3: Estimate Values of States under Policy π

Input: A policy π to be evaluated
Output: Values of $V(s)$, $\forall s$
/* Initialize   */
1 **for** *each state* $s \in S$ **do**
2   Set $V(s)$ to a random value;
3   Set $C(s)$ to 0 ; /* Count of episodes starting at state s */

4 **Loop forever**
5      $s_0 \leftarrow$ a start state for the episode; Pick randomly if there are several start states;
6   Generate an episode following π for H time steps, where H is horizon size:
7          $\langle s_0, a_o, r_o, s_1, a_1, r_1, \cdots s_{H-1}, a_{H-1}, r_{H-1}, s_H \rangle$;
8      $R \leftarrow 0$;
9   **foreach** *step of episode* $: t = H-1, H-2, \cdots$ **do**
10          $R \leftarrow \gamma R + r_t$;
11          $V(s_t) \leftarrow \frac{V(s_t) \times C(s_t) + R}{C(s_t)+1}$;
12          $C(s_t) \leftarrow C(s_t) + 1$

Note that in the inside loop, the values of the states are updated backwards. In addition, in the outer loop, the episode is performed to a horizon H. if the agent gets to a goal state before reaching the horizon, the agent remains in the goal state, looping back to itself (the action performed is *loop back*) with immediate reward of 0.

It is possible to differentiate between two ways of performing the valuations of states in the inner loop. If a state appears more than once in an episode, the agent may update the state's valuation the first time it gets there, a *first-visit algorithm*, or at every visit, called an *every-visit algorithm*. Both approaches lead to convergence, and the choice of one over the other is left to the agent.

However, the every-visit algorithms provide the background that some other more complex reinforcement learning algorithms use.

The algorithms discussed in this section are usually called Monte Carlo (MC) algorithms. Monte Carlo methods are methods that contain a significant random component. They use repeated random experiments or sampling to estimate numerical values. They are used to solve problems when other methods to estimate the values are difficult or impossible. In this case, we are attempting to estimate the values of the states, and there is no simple way to do so in a large space. The Monte Carlo methods we discuss here can potentially perform a large number of random experiments, starting from random start states and learn the $V(s)$ values, for all or a number of states in the search space. Since the MC experiments or episodes start from random (start) states and perform random actions (according to a policy π) in sequence, it is possible that an algorithm is able to obtain values of only the states that are commonly encountered during normal performance of actions and tasks in the environment. This means that because unevaluated or not well-evaluated states usually occur rarely in the search space, and not learning values for them many not affect the general abilities for goal-directed actions in a domain.

5.4.2   Writing the Update Formula Incrementally

In the algorithms above, the update formula for the value of a state s is given as

$$V(s) \leftarrow \frac{V(s) \times C(s) + R}{C(s) + 1} \tag{5.12}$$

Let us rewrite the right-hand side of this assignment.

$$
\begin{aligned}
V(s) &\leftarrow \frac{V(s) \times C(s) + R}{C(s) + 1} \\
&\leftarrow \frac{V(s) \times (C(s) + 1) + R - V(s)}{C(s) + 1} \\
&\leftarrow V(s) + \frac{R - V(s)}{C(s) + 1}
\end{aligned}
\tag{5.13}
$$

If we use a learning rate $\alpha = \frac{1}{C(s)+1}$, we can write

$$V(s) \leftarrow V(s) + \alpha \, (R - V(s)). \tag{5.14}$$

Thus, as a state is visited in an episode, we can update its value using an update formula, which is somewhat similar to formulas we have seen in machine learning earlier. It can be interpreted as follows. We update the value of a state slowly and cautiously, at least as training progresses, so that abrupt changes are avoided as much as possible. The update formula has two parts: i) keep the old value $V(s)$, but ii) change it by a $\Delta V(s)$ update value, which is obtained by computing the difference in cumulative reward R and the current value $V(s)$. Thus, $\Delta V(s)$ may be positive or negative or zero, changing $V(s)$ estimate up or down, or not changing at all, in each visit of the state. The value of $\Delta V(s)$ is tempered by the learning rate α. The learning rate's value is $1, \frac{1}{2} = 0.5, \frac{1}{3} = 0.333, \frac{1}{4} = 0.25, \ldots$, as we have more and more visits to a state. As we can surmise, initially the changes in the values of $V(s)$ may be a abrupt, but the changes become small as we have more and more visits to a state. As the number of visits to a state gets larger, the changes become smaller and smaller. For example, after 100 visits to a state, the value of change is obtained by multiplying $R - V(s)$ by 0.01, making the change small. After an "infinite" number of visits to a state, the value of $V(s)$ does not really change any more, thus converge. Since the value of $V(s)$ changes rapidly early during training, the early episodes are likely to matter more than latter episodes in what value $V(s)$ converges to. However, it is clear that if every state in the space is visited a large number of times, so-called "infinitely often", the values of $V(s)$ will converge for all states s in the space.

Since the update equation, Equation 5.14 is a rewrite of the equation used in Algorithm 5.3, we can stay that Algorithm 5.3 leads to convergence of values of states as well. Since Algorithm 5.1 and Algorithm 5.2 use similar formulation of update for $V(s)$ values, they also result in convergence of $V(s)$ values, as the states are visited "infinitely often".

5.5 Learning Policy: ϵ-Greedy Algorithms

The three algorithms given so far assume that we are following a given policy π, and evaluating the states. Although it is interesting to be able to do so, it does not really solve the problem we set out to address. If the policy is already known, there is really nothing new learning taking place as far as policies go. If the goal is to learn what action to perform when, the agent actually needs to learn the corresponding policy, possibly by performing actions in the real world or a simulated environment. The algorithms we have discussed can be modified to learn policies. In such an endeavor, we can start with a "random" policy and slowly improve upon it, possibly creating a slightly improved policy in each iteration, thus potentially leading to a good, even excellent or expert-level policy after many episodes or iterations.

One way to accomplish it may be the following. We initialize all the $V(s)$ values for all states to 0, or randomly within a certain range with positive as well as negative values. Such an initialization of the values of $V(s)$ implicitly creates a random policy. This is because a policy can be created from the values of the states.

Given a state s, we can look at all states s' reachable from s by performing some action a, and choose to perform the action a' that takes the agent to a state s' using a strategy that leads to performing the "best" action most of the time, but the other non-best actions the remainder of the time. The best action at a certain state a is the one that gives the highest value for the sum of the immediate reward and the valuation of the reached state. The valuation of the reached state is not added directly, but is multiplied by the discount factor γ. Thus, at state a, the "best" action according to the current policy is

$$a = \arg\max_{a'} \left(r(s, a') + \gamma \, V(\delta(s, a')) \right). \tag{5.15}$$

In this formula, $r(s, a')$ is the immediate reward obtained by performing action a' in state s. $\delta(s, a')$ is the state where the agent transitions to by performing action a' in state s. If we write $s' = \delta(s, a')$. i.e., s' is the state that the agent finds itself after performing action a in state s, then $V(\delta(s, a')) = V(s')$ is the valuation of state s'. However, the agent can perform other actions in state s although they are not deemed the best by the current or extant policy. Each such action a' can be evaluated as $r(s, a') + \gamma \, V(\delta(s, a'))$. If we define a valuation function for an action a at a certain state s called $Q(s, a)$, then

$$Q(s, a') = r(s, a') + \gamma \, V(\delta(s, a')).$$

Using this definition, we can write Equation 5.15 as

$$a = \arg\max_{a'} Q(s, a'). \tag{5.16}$$

To learn a policy, we then perform actions following the policy we start with as before, but we do not follow the policy as strictly as the algorithms given earlier. Most of the time, we follow the current policy, but once in a while we randomly perform actions that are different or not the best based on current knowledge. Following the policy strictly will require that the agent always performs the action a in state s that has the highest $Q(s, a)$ value. But, if the agent performs an action a in state s that is not mandated by policy, the agent is exploring a possibility—with the assumption that the current policy or state valuations may be the best learned so far, but there may be a policy that leads the agent to perform a locally non-optimal action at a state s, but it may lead to a better overall policy later because a good policy should lead the agent to obtain high cumulative reward, not just be locally good; it should be globally good.

The algorithm's structure is similar to the prior algorithms we have discussed. The only difference is how the agent generates an episode. In a so-called ϵ-*greedy algorithm*, ϵ being a small positive value between 0 and 1, the agent chooses at any state, a random action, including the "best" action, with probability of ϵ. In addition, if there are $|A(s)|$ actions available at state s, the agent chooses the best action, i.e., the one with the highest $Q(S, a)$ value, with a probability of $1 - \epsilon$, which is high because ϵ is small. For example, if $\epsilon = 0.05$, the "best" action is chosen 95% of the time, and if there are a total of 10 actions (say, including the best action) available in a state, each one of these actions is chosen with a probability of $5/10 = 0.5\%$. In general, since there are $|A(s)|$ actions at state s, each one of these actions can be chosen with a probability of $\frac{\epsilon}{|A(s)|}$, when an action is chosen randomly. Thus, the "best" action is actually chosen with a probability of $1 - \epsilon + \frac{\epsilon}{|A(s)|}$, and all other actions are chosen with a probability of $\frac{\epsilon}{|A(s)|}$. This leads to an algorithm where starting with a randomly initialized vector $\overrightarrow{V}$ of $V(s)$ values, i.e., a random policy, the $V(s)$ values can be improved over time, leading potentially to a much better policy after a while.

Although the algorithm's structure is similar to the prior algorithms, there are some crucial differences. The agent finds a policy by performing actions or simulating actions and getting (simulated) rewards. So, the agent is not beholden to any given policy at any time. The agent initializes all $V(s)$ values randomly. The $V(s)$ values implicitly define a policy since the action to be chosen at a state is the one that has the highest $Q(s, a)$ value. However, the agent does not follow the implicit policy exactly at any time. It performs the action that leads to the highest $Q(s, a)$ value most of the time, following the current implicit policy but deviates from it once in a while, performing a random action. This is called *exploration* since the agent is in pursuit of the highest cumulative reward and the current (greedy) best action may not always be the best choice, considering long-term rewards. Even if the best action is not chosen at a certain time, there may be future actions that may lead to higher cumulative reward for the entire episode or within the specified horizon of the episode.

Once an episode of training has been generated completely or up to a horizon H, the value of the states in the episode are updated either by averaging as before, or by using the incremental update formula discussed earlier. Both approaches are exactly the same.

Generation of episodes and updating of $V(s)$ for the states can continue forever, since reinforcement learning is life-long learning in a sense. However, at any time after a certain episode of training is finished, the agent has an implicitly updated policy, which is keeps on improving under certain conditions of training, as discussed before. After a large number of episodes and/or visits to a state s, its value may converge based upon certain assumptions.

5.5.1 Alternative Way to Learn Policy by Exploitation and Exploration

When the best action as recommended by the current (implicit or explicit) policy is chosen, it is called *exploitation*. Exploitation may not be a great idea for an uninformed or slightly informed agent, but as the agent learns for a while, the estimates of $V(s)$ become better and as a result, exploitation becomes generally a good idea to follow. However, exploration, at least once in a while, is always good for an agent. Instead of taking an ϵ-greedy approach

---

Algorithm 5.4: Learn a New Policy π

Input: A set of states S, and actions that can be performed in each
         state with immediate reward
Output: A Policy π
```
/* Initialize                                                    */
```
1 **for** *each state $s \in S$* **do**
2 $\quad$ Set $V(s)$ to a random value;
3 $\quad$ Set $C(s)$ to 0 ; /* Count of episodes starting at state s */

```
/* Episodes; learning and acting with current learned
   policy can be interspersed                                    */
```
4 **Loop forever**
5 $\quad$ $s_0 \leftarrow$ a start state for the episode; Pick randomly if there are
        several start states;
6 $\quad$ *Generate-ϵ-greedy-episode*$(V(), \epsilon, s_0)$ using π for H time steps,
        where H is horizon size:
7 $\quad\quad$ $\langle s_0, a_o, r_o, s_1, a_1, r_1, \cdots s_{H-1}, a_{H-1}, r_{H-1}, s_H \rangle$;
8 $\quad$ $R \leftarrow 0$;
9 $\quad$ **foreach** *step of episode : $t = H-1, H-2, \cdots$* **do**
10 $\quad\quad$ $R \leftarrow \gamma R + r_t$;
11 $\quad\quad$ $V(s_t) \leftarrow V(s_t) + \alpha \, (R - V(s_t))$ where $\alpha = \frac{1}{C(s_t)+1}$;
12 $\quad\quad$ $C(s_t) \leftarrow C(s_t) + 1$

13

---

14 **Function Generate-ϵ-greedy-episode**$(V(), \epsilon, s_0)$:
15 $\quad$ $s \leftarrow s_0$;
16 $\quad$ **for** $i = 0$ *to* $H-1$ **do**
17 $\quad\quad$ $u \leftarrow$ Generate a random number between 0 and 1;
18 $\quad\quad$ **if** $u \leq \epsilon$ **then**
19 $\quad\quad\quad$ $m \leftarrow$ Generate a random integer between 1 and $|A(s)|$;
20 $\quad\quad\quad$ $a_i \leftarrow A(s_i)(m)$; /* mth possible action in state s_i
                  */
21 $\quad\quad$ **else**
22 $\quad\quad\quad$ $a_i \leftarrow \arg\max_{a'} Q(s, a')$;
23 $\quad\quad$ $r_i \leftarrow r(s_i, a_i)$;
24 $\quad\quad$ $s_{i+1} \leftarrow \delta(s_i, a_i)$
25 $\quad$ **return** $\langle s_0, a_o, r_o, s_1, a_1, r_1, \cdots s_{H-1}, a_{H-1}, r_{H-1}, s_H \rangle$;

26

---

27 **Function Generate-policy**$(V(), S, A)$:
28 $\quad$ $s \leftarrow s_0$;
29 $\quad$ **for** *each state $s \in S$* **do**
30 $\quad\quad$ $\pi(s) \leftarrow \arg\max_{a'} Q(s, a'), \;\; a' \in A(s)$;
31 $\quad$ **return** $\pi()$;

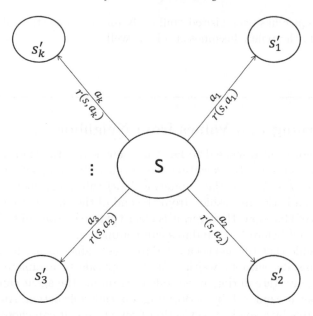

FIGURE 5.6: A state and its neighbors, with actions and rewards

to choose the next action, we can compute the probability by which an action is chosen at a certain state s in another way also.

As shown in Figure 5.6, a state s has neighbors s_1' through s_k' that can be reached by performing an action. These actions form the set $A(s)$ of actions performable in state s. The agent computes the probability that a certain action is chosen, given the current state such that possible actions with higher $Q(s, a) = r(s, a) + \gamma\, V(\delta(s, a))$ values have higher probability of being chosen; but, every possible action has a probability of being chosen, however small. This probability is computed using the so-called softmax function,

$$p\left(a = a_i \mid s\right) = \frac{e^{Q(s, a_i)}}{\sum_{j=1}^{k} e^{Q(s, a_j)}}. \tag{5.17}$$

Higher Q-valued actions have higher probabilities of being chosen.

Once again, the agent can start by initializing all state valuations to random values. Then, the agent generates complete episodes of training or episodes up to a horizon H, following choice of actions by performing exploitation and exploration, based on the currently extant policy, using the softmax formula. After a number of episodes of training the estimates of the values of $V(s)$ start to become stable and converge under the condition that the value of α, the learning parameter, goes down as a state is visited more number of times. As the values of $V(s)$ become more stable, the implicit policy defined by the $V(s)$ values becomes better and more stable. After each state in the space has been visited "infinitely many" times, the values become solidified, i.e., the policy becomes stable as well.

5.6   Learning $V(s)$ Values from Neighbors

In the algorithms discussed so far, the agent learns the values of $V(s)$ for the states in state space by executing an episode of training, either completely or up to a horizon H, and then updating $V(s)$ values considering the entire episode, or in a backward fashion from the end of the episode to earlier states for purposes of efficiency. However, it is clear that an episode of training needs to be completed before $V(s)$ updates can commence.

It is possible that the agent can update $V(s)$ values as it executes actions within an episode, without waiting for the episode to complete (up to the horizon). At any state during an episode of training, the agent can look at all the states that it can reach by performing a permissible action. Among all the reachable states by a single direct action from state s, it can choose an action a that leads to one with the highest value of $Q(s, a) = r(s, a) + \gamma \, V(\delta(s, a))$ where $\delta(s, a)$ is the state that the agent goes to by performing action s in state s.

Algorithm 5.5: Learn a New Policy π

Input: A set of states S, and actions that can be performed in each
        state with immediate reward
Output: A policy π, stating what action is best in which state
```
/* Initialize                                          */
```
1 **for** *each state* $s \in S$ **do**
2   Set $V(s)$ to a random value ;   `/* can be all 0 */`
3 **Loop forever**
4      $s_0 \leftarrow$ a start state for the episode; Pick randomly if there are
        several start states;
5      $s \leftarrow s_0$;
6   **repeat**
7          $a \leftarrow$ Choose an action at state s by combining exploitation
          and exploration using ϵ-greedy or softmax approach;
8          $s' \leftarrow \delta(s, a)$, the state the agent transitions to by performing
          action a in state s;
9          $V(s) \leftarrow r(s, a) + \gamma \, V(s')$;
10        $s \leftarrow s'$
11   **until** H *actions are performed or till a goal state is reached*;

The algorithm looks simpler than some of the algorithms we have discussed so far. The agent performs as many episodes as necessary since reinforcement learning is a life-long learning process. Each episode can be executed till the end, or for a horizon H steps. The agent can choose the next state s' based on the highest $Q(s, a)$ value or a judicious mix of exploitation and exploration as discussed earlier. As an action is performed, the value of the originating state s is updated based on the value of the immediate reward $r(s, a)$ and the discounted value of the resulting state.

Since the $V(s)$ values keep on changing during the training process, the overall policy keeps on changing as well. When the values of $V(s)$ become stable, which is likely to happen after a large number of episodes of training, the agent has learned a stable policy. For values of $V(s)$ to converge, each state s has to be visited infinitely-often, i.e., many times; and, the values of the immediate rewards must be bounded from above by a certain positive constant c.

5.7 Q Learning

So far, we have discussed how to learn valuations $V(s)$ of states , and then to compute a policy based on the values of neighboring states, given a certain state s. We used the valuations of actions in states, $Q(s, a)$, to help evaluate states in some algorithms. $\pi(s)$ can simply be the action that takes the agent to the highest valued neighbor. There is an algorithm called Q-learning that uses $Q(s, a)$ values directly, without computing $V(s)$ values. As we have noted earlier, a $Q(s, a)$ value estimates the goodness of action a when performed in state s, no matter at what time. The Q-learning algorithm has been one of the most widely discussed and used algorithms in reinforcement learning.

The Q-learning algorithm is simple and is similar to the last V-learning algorithm. It maintains a table called the Q-table that keeps the running values for each $\langle s, a \rangle$ pair. The $Q(s, a)$ table (see Figure 5.7 is initialized to some small random values or all zeros. The $Q(s, a)$ values are updated as episodes are performed continuously. As the agent performs an action a in state s, it updates the $Q(s, a)$ value for the $\langle s, a \rangle$ combination. At any time, based on the $Q(s, a)$ values, we can compute the values of states as

$$V(s) = max_a \, Q(s, a). \tag{5.18}$$

The agent learns by performing actions in episodes, like all reinforcement learning algorithms discussed. In each episode, the agent performs an action following a strategy that mixes exploration and exploitation. Once an agent performs an action a in state s, it updates the corresponding $Q(s, a)$ value,

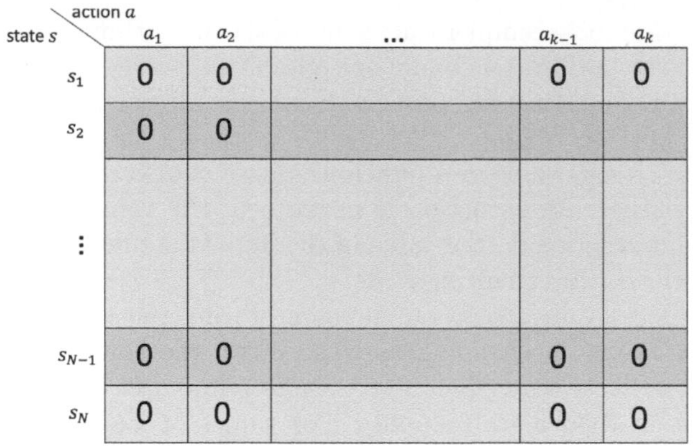

state s \ action a	a_1	a_2	...	a_{k-1}	a_k
s_1	0	0		0	0
s_2	0	0			
⋮					
s_{N-1}	0	0		0	0
s_N	0	0		0	0

FIGURE 5.7: The $Q(s, a)$ Table Initalization

using the formula

$$Q(s, a) \leftarrow r(s, a) + \gamma \, max_{a'} Q(\delta(s, a), a'). \tag{5.19}$$

In other words, the new $Q(s, a)$ value for action a, performed in state s is obtained by updating the old value by a change component, obtained by looking at the state the agent finds itself after performing action a. Let the state the agent finds itself after performing action a in state s be $s' = \delta(s, a)$. In state s', the agent can possibly perform several actions. Of these actions in state s', the agent considers the value of the highest valued action, dampens in by the discount factor and uses the resulting value as the update value for $Q(s, a)$.

For the Q-learning algorithm to end, the $Q(s, a)$ values should converge for every $\langle s, a \rangle$ pair. This happens if each $\langle s, a \rangle$ pair is visited " infinitely-often" or a large number of times. As mentioned earlier, this happens if the values of the immediate rewards are bounded from above by a positive constant c. In other words, the immediate reward values should not be very large.

5.8　Reinforcement Learning in R

There are several libraries in R that implement reinforcement learning, although the support is not as strong as in Python. The reinforcement learning package used here is called **reinforcelearn**. It can be downloaded from github using devtools.

```
devtools::install_github("markusdumke/reinforcelearn")
```

Algorithm 5.6: Learn a New Policy π

Input: A set of states S, and actions that can be performed in each
state with immediate reward

Output: A policy π, stating what action is best in which state

```
/* Initialize                                              */
```
1 **for** *each state $s \in S$* **do**
2   ⌊ Set $V(s)$ to a random value ;   `/* can be all 0 */`
3 **Loop forever**
4    $s_0 \leftarrow$ a start state for the episode; Pick randomly if there are
several start states;
5    $s \leftarrow s_0$;
6   **repeat**
7       $a \leftarrow$ Choose an action at state s by combining exploitation
and exploration using ϵ-greedy or softmax approach;
8       $s' \leftarrow \delta(s, a)$, the state the agent transitions to by performing
action a in state s;
9       $Q(s, a) \leftarrow r(s, a) + \gamma \, max_{a'} Q(s', a')$;
10       $s \leftarrow s'$
11   **until** *H actions are performed or till a goal state is reached*;

The library provides functions to create environments, agents, learning algorithms, and finally produce interactions between the agent and the environment, with or without learning. The library provides several builtin environment types such as Markov Decision Processes, Gridworlds, and more general environments.

5.8.1   Q-Learning in a Simple 3×3 Gridworld

Below, we develop a simple gridworld environment with a simple reward structure, and then use it to learn a policy using Q-learning.

```
1  #file QLearning3x3.R
2
3  library ("reinforcelearn")
4
5  #Create a  3x3 gridworld environment with initial state numbered 0 and
6  #final state numbered 8, with a discount factor of 0.9
7  env <- makeEnvironment("gridworld", shape = c(3, 3), initial.state = 0L,
8        goal.states = 8L, discount = 0.9 )
9
10 env$visualize()
11
12 #Create the sparse rewards matrix and link it to the environment
13 #reinforcelearn uses indices starting from 0 whereas
14 #R uses indices starting from 1. One needs to be careful!
15 RewardMatrix <- matrix(0, nrow=9, ncol=4) #All immediate rewards are 0
```

```
16  RewardMatrix[6,4] <- 100 #row 6 in R is   state 5 in reinforcelearn
17  RewardMatrix[8,2] <- 100 #row 8 in R is state 7 in reinforcelearn
18  env$rewards <- RewardMatrix
19
20  #Create the qlearning agent
21  agent <- makeAgent(policy = "softmax", val.fun = "table",
22    algorithm = "qlearning")
23
24  #######Inspect Q(s,a) and V(s) values before learning
25  #Get the Q (s,a) values learned
26  getValueFunction(agent)
27  #Print the state values learned
28  getStateValues(getValueFunction(agent))
29
30  ###################
31  #Interact with the environment over 1 episode only and see inspect
32  #Q(s,a) and V(s) values
33  interact(env, agent, n.episodes = 1, learn = TRUE, visualize = TRUE)
34  #Get the Q (s,a) values learned
35  getValueFunction(agent)
36  #Print the state values learned
37  getStateValues(getValueFunction(agent) )
38
39  ###################
40  #Interact with the environment over 10 more episodes and see inspect
41  #Q(s,a) and V(s) values
42  interact(env, agent, n.episodes = 10, learn = TRUE)
43  #Get the Q (s,a) values learned
44  getValueFunction(agent)
45  #Print the state values learned
46  getStateValues(getValueFunction(agent) )
47
48
49  ###################
50  #Interact with the environment over 100 more episodes and see inspect
51  #values Q(s,a) and States. Learning should be finished by now.
52  interact(env, agent, n.episodes = 1000, learn = TRUE)
53  #Get the Q (s,a) values learned
54  getValueFunction(agent)
55  #Print the state values learned
56  getStateValues(getValueFunction(agent) )
57
58  ###########
59  #Visualize the agent playing; it plays the same way every time!
60  for (i in 1:3) {
61    env$reset()
62    env$visualize()
63    # comment in next line to wait on enter before taking next action.
64    # invisible(readline(prompt = "Press [enter] to take the next action"))
65    interact(env, agent, n.episodes = 1, learn = FALSE, visualize = TRUE)
66    Sys.sleep (1) #Sleep for a minute at the end of an episode
67  }
```

Line 3 makes the `reinforcelearn` library available to the program, assuming it has been installed. Lines 5-8 create an environment using the `makeEnvironment` function. The first argument is a string indicating that a special type of environment called the `gridworld` is created. The other environments possible are `mdp`, `gym`, and `custom`, which stand for a Markov Decision Process, the OpenAI Gym and a custom-made environment. A gridworld environment is simply a rectangular space with numbered cells. Movement is allowed through actions of going left, right, up and down: numbered 1, 2, 3 and 4. This set of actions is created automatically by the `makeEnvironment` function. The cells are numbered starting from 0, row by row. In this specific case, the initial state for the environment is the cell numbered 0, the leftmost cell of the top row, and the goal state is the state numbered 8, the rightmost cell of row 3. The environment can be visualized as below with the agent sitting in state 0 (line 10).

```
o - -
- - -
- - -
```

The assumption here is that the environment gives the rewards. The agent will learn to start from the initial state and travel through the grid cells to the goal state. `makeEnvironment` allows the specification of only one initial state, but one or more goal states. The agent initially does not know how to go from the initial state to the final state in an optimal way. It learns to do so using a learning algorithm that uses a discount factor of 0.9.

The default setting in the `reinforcelearn` library sets the immediate rewards for getting to the goal state from any other state 0. It also sets all other immediate rewards to -1. However, we want to change these values to what we want as described next. Lines 12-16 create a matrix of immediate rewards called `RewardMatrix`. It is a sparse matrix, initially all set to 0. Lines 14 and 15 set the immediate reward values for going left from state 7 to 8, and for going don from state 5 to state 8 to 100. These are the only two non-zero immediate rewards, in the spirit of keeping the immediate rewards "rare" or infrequent. Line 16 sets the rewards associated with the environment `env` to the `RewardMatrix` created separately. The immediate rewards are seen below.

```
> env$rewards
      [,1] [,2] [,3] [,4]
[1,]    0    0    0    0
[2,]    0    0    0    0
[3,]    0    0    0    0
[4,]    0    0    0    0
[5,]    0    0    0  100
[6,]    0    0    0    0
[7,]    0  100    0    0
[8,]    0    0    0    0
[9,]    0    0    0    0
```

The states and the immediate rewards based on specific actions performed in specific states are also shown in Figure 5.8.

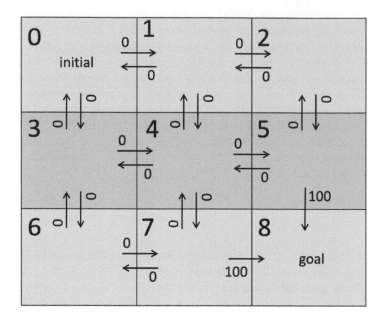

FIGURE 5.8: Simple 3 × 3 Rectangular Space with Immediate Rewards

Once we have created an environment, the next step is to create an agent. Lines 18-20 create an agent that uses `softmax` approach to choosing actions among the possible ones in an environment. In other words, it converts the $Q(s, a)$ values for all possible actions to probabilities using the softmax function as discussed earlier, and chooses one of the actions with the probabilities computed. Thus, it uses a combination of exploitation and exploration in choosing the next action in any state. The possible policy values are `greedy` for pure exploitation, and `epsilon.greedy` discussed earlier. The agent also stores the $Q(s, a)$ values in a dynamic programming table as discussed earlier. The agent created uses Q-learning to learn by performing episodic sequences of actions in the environment.

Before any learning, we can try to print the $Q(s, a)$ values and the state values (lines 24-28) , and get the following.

```
> getValueFunction(agent)
Error in getValueFunction(agent) :
  No value function weights found in the agent object.
> getStateValues(getValueFunction(agent))
Error in getValueFunction(agent) :
  No value function weights found in the agent object.
```

To test how the Q-values and state values change after just one interaction of the agent with the environment, line 33 causes the agent to perform just one episode of training with visualization, and then line 35 and 37 print the values. Since the agent starts with a blank slate as far valuation of actions and states, the episode may take a long time, sometimes tens or even hundreds of steps. However, the episode changes the Q-value and state values each in one location.

```
> getValueFunction(agent)
> #Get the Q (s,a) values learned
> getValueFunction(agent)
        [,1] [,2] [,3] [,4]
[1,]     0    0    0    0
[2,]     0    0    0    0
[3,]     0    0    0    0
[4,]     0    0    0    0
[5,]     0    0    0    0
[6,]     0    0    0    0
[7,]     0    0    0    0
[8,]     0   10    0    0
[9,]     0    0    0    0
> #Print the state values learned
> getStateValues(getValueFunction(agent) )
[1]  0  0  0  0  0  0  0 10  0
```

Line 42 causes the agent to interact with the environment for an additional ten episodes. The agent learns very quickly, as is clear from the output of R. In addition, the Q-values and state values (lines 44 and 46) also have changed substantially.

```
$returns
 > getValueFunction(agent)
            [,1]            [,2]        [,3]        [,4]
[1,] 0.02987253  0.07467921 0.057644839 0.5363527
[2,] 0.01749404  0.00000000 0.013851000 0.6499918
[3,] 0.00000000  0.00000000 0.000000000 0.0000000
[4,] 0.02631690  0.00000000 0.003490452 4.1405101
[5,] 0.00000000  0.00000000 0.000000000 9.6501230
[6,] 0.00000000  0.00000000 0.000000000 0.0000000
[7,] 0.00000000 21.69346312 0.000000000 0.0000000
[8,] 0.00000000 68.61894039 0.000000000 0.0000000
[9,] 0.00000000  0.00000000 0.000000000 0.0000000
> getStateValues(getValueFunction(agent) )
[1]   0.5363527  0.6499918  0.0000000  4.1405101  9.6501230  0.0000000 21.6934631
[8] 68.6189404  0.0000000
```

Line 51 causes the agent to interact with the environment another 50 times, learning at the same time, using Q-learning. The output shows that the agent learns very quickly, and after another four episodes, the cumulative return converges, and the number of steps converges to the optimal 4.

```
$returns
 [1] 59.0490 72.9000 53.1441 72.9000 72.9000 72.9000 72.9000 72.9000 72.9000
[10] 72.9000 72.9000 72.9000 72.9000 72.9000 72.9000 72.9000 72.9000 72.9000
[19] 72.9000 72.9000 72.9000 72.9000 72.9000 72.9000 72.9000 72.9000 72.9000
[28] 72.9000 72.9000 72.9000 72.9000 72.9000 72.9000 72.9000 72.9000 72.9000
[37] 72.9000 72.9000 72.9000 72.9000 72.9000 72.9000 72.9000 72.9000 72.9000
[46] 72.9000 72.9000 72.9000 72.9000 72.9000
```

```
$steps
 [1] 6 4 7 4 4 4 4 4 4 4 4 4 4 4 4 4 4 4 4 4 4 4 4 4 4 4 4 4 4 4 4 4 4 4 4 4 4 4
[39] 4 4 4 4 4 4 4 4 4 4 4
```

Lines 53 and 55 print the Q-values and state values again. These happen to be final values.

```
> #Get the Q (s,a) values learned
> getValueFunction(agent)
            [,1]        [,2]         [,3]         [,4]
[1,] 0.10386798   0.2439548 0.192959453 62.336668
[2,] 0.01749404   0.0000000 0.013851000  2.645627
[3,] 0.05849926   0.0000000 0.000000000  0.000000
[4,] 0.02631690   0.0000000 0.003490452 76.575701
[5,] 0.00000000   0.0000000 0.000000000 20.521714
[6,] 0.00000000   0.0000000 0.000000000  0.000000
[7,] 0.00000000  88.7836875 0.000000000  0.000000
[8,] 0.00000000  99.8382691 0.000000000  0.000000
[9,] 0.00000000   0.0000000 0.000000000  0.000000
> #Print the state values learned
> getStateValues(getValueFunction(agent) )
[1] 62.33666766   2.64562669   0.05849926 76.57570121 20.52171362   0.00000000
[7] 88.78368751  99.83826907   0.00000000
```

This is a very simple environment, and it takes the Q-learning agent to learn the policy to go from the initial state to the goal state in 15 steps or so. In general, for more complex environments and sets of actions, the learning is much slower. When the learned agent is asked to perform an episode (lines 58-66) without learning, based on the current knowledge it has acquired, the moves it makes are shown below.

```
o  -  -
-  -  -
-  -  -


-  o  -
-  -  -
-  -  -


-  -  o
-  -  -
-  -  -


-  -  -
-  -  o
-  -  -


-  -  -
-  -  -
-  -  o
```

It is clear that the agent has learned to move from the initial state 0 to the goal state 8 without any dithering in an optimal manner. The program asks it to perform a single episode three times, and each time it performs the exact same steps.

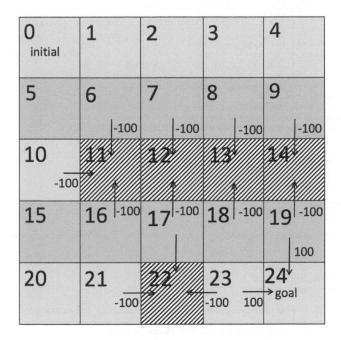

FIGURE 5.9: A Simple 5×5 Maze

5.8.1.1 Q-Learning in a Simple 5×5 Maze

In this example, the environment is more complex than the previous example, with a total of 25 states in a 5×5 square configuration. The initial state is 0—the left top state, and the goal state is 24—the right bottom state. The environment has some barriers as shown in Figure 5.9. The immediate rewards for transition to the final state are 100 from the neighboring states. Thus, the immediate reward for down action in state 19 is 100, and so is the immediate reward for left action in state 23. The barriers are modeled in the environment by in making the immediate rewards -100. Thus, the immediate reward for right action in state 10 is -100. The immediate rewards for down actions in states 6, 7, 8 and 9 are -100, as well as for up actions in states 16, 17, 18 and 19. These immediate rewards are shown in Figure 5.9. The immediate rewards not shown are all 0.

```
1   #file QLearning5x5.R
2
3   library ("reinforcelearn")
4   #Create a  5x5 gridworld environment with initial state numbered 0 and
5   #final state numbered 8, with a discount factor of 0.9
6   env <- makeEnvironment("gridworld", shape = c(5, 5), initial.state = 0,
7          goal.states = 24,   discount = 0.9 )
8   env$visualize()
9
10  #Create locations of barriers with rewards and link to the environment
11  #reinforcelearn uses indices starting from 0 whereas
12  #R uses indices starting from 1. One needs to be careful!
```

```
13  RewardMatrix <- matrix(0, nrow=25, ncol=4) #All immediate rewards are 0
14
15  #All these rewards are high negative
16  BarrierLocations <- matrix (c (11, 2, 7, 4, 8, 4, 9, 4, 10, 4, 22, 2,
17          24, 1, 17, 3, 18, 3, 19, 3, 20, 3, 18, 4),
18          ncol = 2, byrow = TRUE)
19  #Write code to set them to all REWARD.AT.BARRIER
20  REWARD.AT.BARRIER = -100
21  for (row in 1:nrow (BarrierLocations)){
22      RewardMatrix [BarrierLocations[row,1],
23      BarrierLocations [row,2]] <- REWARD.AT.BARRIER
24  }
25
26  #Add positive rewards at goal
27  RewardMatrix [20, 4] <- 100
28  RewardMatrix [24, 2] <- 100
29
30  #Link RewardMatrix to the environment
31  env$rewards <- RewardMatrix
32
33  #Display the immediate rewards and check to make sure they are correct
34  env$rewards
35
36  #Create the qlearning agent
37  agent <- makeAgent(policy = "softmax", val.fun = "table",
38      algorithm = "qlearning")
39
40  #Interact with the environment over a large number of episodes and inspect
41  #values Q(s,a) and States. Learning should be finished by now.
42  interact(env, agent, n.episodes = 100, learn = TRUE)
43  #Get the Q (s,a) values learned
44  getValueFunction(agent)
45  #Print the state values learned
46  getStateValues(getValueFunction(agent) )
47
48  ##########
49  #Visualize the agent playing
50  env$reset()
51  env$visualize()
52  interact(env, agent, n.episodes = 1, learn = FALSE, visualize = TRUE)
```

Lines 6-8 create the maze environment, with a discount factor of 0.9 for reinforcement learning. The reinforcelearn library's makeEnvironment function creates the immediate rewards for all actions to be -1 by default and 0 for actions to the goal state. The program overwrites these default rewards by creating a RewardMatrix (line 13), where all initial values are set to 0. At the barriers, for the actions mentioned earlier are set to -100 (lines 16-24). Lines 27-28 set the immediate rewards to 100 for the actions that lead to the goal state. Line 31 lines these new immediate rewards to the environment env. Line 34 displays the immediate rewards for checking.

```
> env$rewards
        [,1] [,2] [,3] [,4]
  [1,]    0    0    0     0
  [2,]    0    0    0     0
  [3,]    0    0    0     0
  [4,]    0    0    0     0
  [5,]    0    0    0     0
  [6,]    0    0    0     0
  [7,]    0    0    0  -100
```

```
[8, ]      0      0      0 -100
[9, ]      0      0      0 -100
[10, ]     0      0      0 -100
[11, ]     0   -100      0      0
[12, ]     0      0      0      0
[13, ]     0      0      0      0
[14, ]     0      0      0      0
[15, ]     0      0      0      0
[16, ]     0      0      0      0
[17, ]     0      0   -100      0
[18, ]     0      0   -100  -100
[19, ]     0      0   -100      0
[20, ]     0      0   -100    100
[21, ]     0      0      0      0
[22, ]     0   -100      0      0
[23, ]     0      0      0      0
[24, ] -100    100      0      0
[25, ]     0      0      0      0
```

Line 37 creates a Q-learning agent using softmax for choosing an action at every step. The use of softmax leads to an agent that chooses actions in an exploratory manner. Line 42 causes the agent to perform Q-learning for 100 episodes, with lines 44 and 46 printing the $Q(s, a)$ values and the values for states respectively. One of the outputs of the interact function shows the following.

```
$steps
 [1] 322 121  67  45 151 408  96  71 144  54 212 123  12  15 252  43  15  60
[19]  68  58  64  21  22  80   9  14   8  52  17   9  20  35  63  19   9   8
[37]   8   9   8   9   8   8   8   8   8   8   8   8   8   8   8   8   8   8
[55]   8   8   8   8   8   8   8   8   8   8   8   8   8   8   8   8   8   8
[73]   8   8   8   8   8   8   8   8   8   8   8   8   8   8   8   8   8   8
[91]   8   8   8   8   8   8   8   8   8   8
```

This shows that by episode 41, the number of steps in the episodes has stabilized to 8, and the agent has learned how to navigate the maze. Lines 50-52 show how the agent navigates the maze with the knowledge it has acquired via reinforcement learning.

```
o - - - -
- - - - -
- - - - -
- - - - -
- - - - -

- - - - -
o - - - -
- - - - -
- - - - -
- - - - -

- - - - -
```

```
 -  -  -  -  -
 o  -  -  -  -
 -  -  -  -  -
 -  -  -  -  -

 -  -  -  -  -
 -  -  -  -  -
 -  -  -  -  -
 o  -  -  -  -
 -  -  -  -  -

 -  -  -  -  -
 -  -  -  -  -
 -  -  -  -  -
 -  o  -  -  -
 -  -  -  -  -

 -  -  -  -  -
 -  -  -  -  -
 -  -  -  -  -
 -  -  o  -  -
 -  -  -  -  -

 -  -  -  -  -
 -  -  -  -  -
 -  -  -  -  -
 -  -  -  o  -
 -  -  -  -  -

 -  -  -  -  -
 -  -  -  -  -
 -  -  -  -  -
 -  -  -  -  o
 -  -  -  -  -

 -  -  -  -  -
 -  -  -  -  -
 -  -  -  -  -
 -  -  -  -  -
 -  -  -  -  o
```

5.9 Conclusions

This chapter introduced the topic of reinforcement learning along with a number of examples, and followed it by the development of a number of algorithms for reinforcement learning, culminating in a discussion of Q-learning. Reinforcement learning is different from both supervised and unsupervised learning, and occupies a place that has elements of both and more. It is used by agents that learn by performing tasks in an environment by receiving immediate rewards—positive as well as negative. The chapter then presented a couple of examples where reinforcement learning problems were solved using the `reinforcelearn` package in R. For those who are interested in working with OpenAI's Gym environment, the `reinforcelearn` package provides an environment to do so. The package also provides an easy way to incorporate artificial neural networks in reinforcement learning. These are left for the student to learn on his or her own. For those who want to solve sophisticated problems using deep reinforcement learning, Python provides a large number of tools.

Exercises

1. The reinforcement learning environments discussed in this chapter are all deterministic. Assume the environment is a rectangular grid where the agent moves from an initial state to a goal state, using actions of going left, right, up and down. We do not want the agent to fall off the grid at the edges on all four sides: north, south, east and west. How can this requirement be represented explicitly in the description of the environment? Are there alternatives that can achieve the same purpose?

2. In a deterministic grid environment, suppose there are barriers like in one of the examples shown in this chapter. The example modeled barriers with a number of states, with high negative immediate rewards if an agent wants to move to a state that is a part of the barrier region. How do you decide on the amount of negative reward? Can you think of any other ways to model barriers?

3. In a deterministic grid environment, usually the immediate rewards are sparse. In other words, the percentage of actions that get non-zero rewards in certain states among all possible state-action pairs is usually very small. Do you think there is a relationship between the percentage of non-zero immediate rewards and the speed of learning? Explain why or why not.

4. The environment in reinforcement learning can be non-deterministic. In a non-deterministic environment, the transitions due to actions may be stochastic, i.e., one or more actions may lead to different target states with different probabilities. How can one express stochastic state

transitions? How can it be incorporated in the learning of values for states? How can it be incorporated in the learning of values of actions in states?

5. In a non-deterministic grid environment, the rewards obtained by performing an action in one or more states may be non-deterministic as well. In this scenario, when an agent performs a certain action in a state, the rewards can vary with probability values associated with various rewards. How can on express the stochastic nature of rewards?

6. Whether it is a deterministic or non-deterministic environment, a reinforcement learning agent needs to learn the values of states in the environment, or equivalently learn the values of actions in the context of states. In Q-learning, the table update happens by looking ahead one time step. It is possible that the $Q(s, a)$-values will be learned better if the agent could look several steps into the future. How can an agent incorporate the process of looking ahead several steps in reinforcement learning?

7. How would you represent a tic-tac-toe board as an environment to a reinforcement learning agent? How many states are possible in this game? How would you represent moves and immediate rewards?

8. In the game of chess, the number of states is very large, approximately 10^{120} by some accounts. Creating a vector for state-value learning, or a table for $Q(s, a)$-value learning is impossible with such a large number of states since we cannot enumerate all the states in a program. We also know that it is possible to write reinforcement learning programs for learning how to play chess. How do such programs keep track of such a large number of states?

9. If the number of states and actions is "large", the learning agent may decide to save some of the interactions in memory so that the agent can train on these interactions from time to time. Do you think training on saved interactions will help an agent learn faster? Why or how? What kinds of interactions should be chosen if such memory replay is helpful? How many interactions is likely to be stored for memory replay?

10. The Q-learning algorithm learns values of actions in the context of states in which they are performed. If the number of states is large and/or actions is large, it may not be possible to store a table of $Q(s, a)$-values. However, the fundamental goal of Q-learning is to learn what action to perform in which state, i.e., a so-called policy. In other words, the agent needs to learn a function that "predicts" the action to perform given the representation of the current state. How can a programmed agent attempt to predict the function to perform given a state, in this scenario?

Chapter 6

Unsupervised Learning

This chapter discusses unsupervised learning. In unsupervised learning, there is no labeled data for training. To produce reliable labels on data examples as required by supervised learning, someone must have spent time and other resources. For example, to label images as cats, dogs and tigers, someone must have looked at each image and labeled it with the correct type. To label a million images each with one of a thousand possible different types, it is likely that initially an already existing classifier that is fairly good is used to perform classification and label the images. However, following the initial automatic labeling, one or more individuals must have gone through the images, checking on the labels to make sure they are correct. When a large number of items are to be labeled, often previously developed classifiers are used. In addition, often a number of individuals are recruited in a collective manner (as in Amazon Mechanical Turk[1]) to ensure correctness. Sometimes, labels may be checked randomly, if the number of examples is very large. However, in such cases, care is taken to ensure that the labels are correct with a very high probability, say 99%. The essence of this discussion is that labeling requires resources such as the involvement of existing classifiers and human labelers. These are considered "supervisors", and as a result, the learning using such data is called *supervised learning*. The two predominant forms of supervised learning are regression and classification, discussed in earlier chapters fo this book.

In contrast to supervised learning, unsupervised learning does not have access to labeled data. It takes an unlabeled dataset and attempts to find patterns within it. There are several forms of unsupervised learning such as clustering, association mining, outlier detection, feature selection or extraction, and computing so-called embeddings. The process of clustering attempts to group the dataset such that the examples in a group are as similar as possible, and examples in different groups are as dissimilar as possible. In contrast, given a dataset, outlier mining attempts to find data examples that do not fit well in any groups that may be present in the data. Given a dataset, feature selection attempts to find features or attributes that are essential for the task at hand, supervised or unsupervised. Given a so-called market-basket dataset, association rule mining finds associations or co-occurrence relations among individual data items. In simple terms, a market-basket dataset contains examples such as those in shopping transactions, where an example is a

---

[1] https://www.mturk.com

DOI: 10.1201/9781003002611-6

set of items bought together in a "basket". Given a large number of transactions, association mining finds high fidelity co-occurrence relationships among items, e.g., what subset of items occur frequently, and assuming certain items or items have been "bought" together or have co-occurred, which other items are highly likely to co-occur. The idea of embedding is fairly new. It usually refers to embedding a discrete item in a high-dimensional space. For example, given a large body or corpus of textual documents, it is possible to obtain co-occurrence relationships taking into account the propensity of individual words to co-occur with certain other words, and use the information to obtain a vector representation of the word in a high-dimensional space. This high-dimensional vector representation of a discrete word is called *embedding*. It is possible to obtain such embeddings for not only words in text, but also for nodes in a graph in various domains, and for chemical elements in a large number of chemical formulas, among others.

This chapter discussed the most common topic in unsupervised machine learning—clustering.

6.1　Clustering

Given a number of unlabeled examples, the goal of clustering is to find so-called natural groupings of the elements. The groupings are found in such a way that similar items are grouped together and the dissimilar items are grouped separately, as optimally as possible. Thus, to be able to cluster, it is essential to have a measure for computing how similar or dissimilar two items are, or two groups of items are. This is also called a distance metric. The most common metric to compute distance between two individual objects is the Euclidean distance, which is well-known and easily computed. Given two data examples, $\vec{x}^{(i)}$ and $\vec{x}^{(j)}$ each with n components, the Euclidean distance is computed as

$$\left(\vec{x}^{(i)}, \vec{x}^{(j)} \right) = \sqrt{\left(\vec{x}_1^{(i)} - \vec{x}_1^{(j)} \right)^2 + \cdots + \left(\vec{x}_n^{(i)} - \vec{x}_n^{(j)} \right)^2} \qquad (6.1)$$

$$= \sqrt{\sum_{l=1}^{n} \left(\vec{x}_n^{(i)} - \vec{x}_n^{(j)} \right)^2}.$$

There is a large variety of clustering algorithms. This is primarily because it is difficult to clearly define the concept of a cluster. The definition usually depends on the domain of application, or the problem to be solved. In this chapter, we discuss three types of clustering algorithms. Under each type, many algorithms have been developed. We discuss only one example algorithm of each type. The types of algorithms we discuss are centroid-based algorithms, hierarchical algorithms and density-based algorithms.

A centroid-based algorithm represents a cluster, i.e., the subset of points that belong to the cluster with a so-called centroid, also called representative or prototype. The prototype is usually computed as having the average values of each feature that the data points are used to describe. In other words, the algorithm computes the average of each feature separately, and uses the vector of the averages as the prototype for the cluster. In such a case, the prototype may not be a real data point. There are some centroid-based clustering algorithms that force the prototypes to be real points.

A hierarchical clustering algorithm builds clusters in a hierarchical manner. There are two sub-types. Divisive algorithms start with the entire dataset as a single cluster, and break it up into two or more subsets or clusters using a distance-based (or, similarity-based) criterion. It continues to divide the dataset into smaller clusters recursively. The algorithm stops the process of dividing based on a stopping criterion, and the clusters present at the time (at the lowest level) are considered to be the found clusters. In contrast, agglomerative clustering starts by making a cluster out of each data example, and then combines two clusters at a time recursively based on a distance or similarity criterion. It continues to merge two clusters at a time till a stopping criterion is reached.

Both centroid-based clustering and hierarchical clustering use the concept of distance or similarity between two points, or between clusters, or between clusters and an example. A centroid-based clustering algorithm like K-means usually requires, as given, the number of clusters to be find. It is quite likely that the number of clusters is unknown, and has to be guessed and several values tried. In hierarchical clustering, the number of clusters does not need to be known a-priori. However, hierarchical clustering still needs some metric for measuring distance or similarity. Usually, Euclidean distance or a variant of it is used. It also needs a parameter to specify the stopping criterion.

The shape of clusters obtained by a centroid-based clustering algorithm like K-means is spherical or ellipsoidal. Such an algorithm cannot find clusters of other shapes. In hierarchical clustering, whether divisive or agglomerative, the cluster shapes depend on the distance metric used. Assuming the distance metric is Euclidean or a variant, the clusters obtained at any stage are spherical or ellipsoidal, just like centroid-based clustering. However, in reality, not all clusters in all domains or problems are likely to be spherical or ellipsoidal or even convex, and can be concave including ring-shaped or very odd shaped. To find such clusters, clustering algorithms such as density-based clustering have been introduced. In density-based clustering, a distance metric is not used. Instead, for each point, the algorithm computes what is called its *density* as the number of points in a small ϵ-sized neighborhood of it. Points of similar density in a connected neighborhood form a single cluster. Points of similar or dissimilar densities that are separated by non-dense or sparse regions form different clusters. Clusters can be of arbitrary shape. Some points in non-dense regions may be outliers as well.

6.2   Centroid-Based Clustering

As mentioned earlier, in centroid-based clustering, each cluster in the set of clusters produced is assumed to have a central or focal point, called its mean or prototype or centroid. A simple prototype of a cluster can be computed as the mean of all the points in the cluster. This works only if all the features of a data point are numerical and averages can be computed. Otherwise, the centroid or prototype needs to be computed differently. In this chapter, we discuss the most common centroid-based algorithm called the K-means algorithm.

6.2.1   K-Means Clustering

K-means clustering assumes that all features of the items are numeric so that Euclidean distance can be used as distance metric. It is an iterative algorithm that needs as input a positive integer $K \geq 1$, which instructs the algorithm the number of clusters to look for. For example, if $K = 5$, the algorithm finds 5 groups in the data.

The K-means algorithm starts by intializing K means. K is a parameter for the algorithm, given by the user, possibly based on some initial analysis of the data. Each one is called a mean because it is the mean or average of the number of associated points. Each mean can be thought of as a representative or prototype of the data points associated with it.

The goal of the K-means clustering algorithm is to group the input points into K regions such that points in each region are maximally similar, and at the same time, maximally dissimilar from the sets of points in the other clusters. As far as the algorithm goes, there is no mean to start with; and, hence, the initial K means are randomly chosen. The means refined iteratively until a better set of means cannot be found. The means, at this stage, are the final means, and the sets of points associated with each point are considered a cluster. Since, through each iteration, the algorithm always maintains K means, there are K final clusters. The K-means algorithm is given as Algorithm 6.1.

The *repeat* loop in the algorithm can be terminated in two ways. The algorithms can run till the set of means does not change very much, or the loop runs a certain number of pre-specified times T. Running for a certain given number of times can be useful if convergence is slow or there is no convergence. As for convergence, one approach may be to look at the difference between the current and previous sets of means and stop if the difference is below a certain threshold. If $\mathbf{M}^{(old)} = \left\{ \vec{m}_1^{(old)}, \cdots \vec{m}_k^{(old)}, \cdots \vec{m}_K^{(old)} \right\}$ is the old set of means

---

Algorithm 6.1: K-Means Algorithm

Input: A dataset of unlabeled points $\mathbf{D} = \{\vec{x}_1, \cdots, \vec{x}_i, \cdots, \vec{x}_N\}$
Input: K, a user-specified number of clusters to be found in the
      dataset $\mathbf{D}$, a stopping parameter α
Output: K prototype points for K clusters, with each point in $\mathbf{D}$
       associated with a specific prototype

```
/* Initialize                                              */
```
1   $\mathbf{M} \leftarrow$ Randomly initialize each of the K means
    $= \{\vec{m}_1, \cdots, \vec{m}_k, \cdots, \vec{m}_K\}$ to a distinct data point
2   **repeat**
3   //Associate each data point with a mean or prototype
4   **for** $i = 1$ **to** N **do**
5        $\lfloor$ $prototype_i \leftarrow \arg\min_{k=1 \ldots K} \; d\left(\vec{x}_i, \vec{m}_k\right)$
6   //Compute new means, Set each mean to 0 so we can compute
         the new means by iterating over all of N points
7      $\vec{m}_i \leftarrow 0, i = 1 \cdots K$
8      $npoints_i \leftarrow 0, i = 1 \cdots K$
9   **for** $i = 1$ **to** N **do**
10       $\vec{m}_{prototype_i} \leftarrow \dfrac{npoints_{prototype_i} \times \vec{m}_{prototype_i} + \vec{x}_i}{npoints_{prototype_i} + 1}$
11       $\lfloor$ $npoints_{prototype_i} \leftarrow npoints_{prototype_i} + 1$
12   **until** $\mathbf{M}$ stabilizes or for a certain number of iterations T using
    Equation *6.2*;
13   //Cluster C_k contains points that have $prototype_i = \vec{m}_k, \forall i, \forall k$
14   **return** $\mathbf{M} = \{\vec{m}_1, \cdots, \vec{m}_k, \cdots, \vec{m}_K\}$

---

and $\mathbf{M}^{(new)} = \left\{\vec{m}_1^{(new)}, \cdots \vec{m}_k^{(new)}, \cdots \vec{m}_K^{(new)}\right\}$ is the new set of means, the difference can be computed as

$$\mathbf{\Delta M} = \sum_{k=1,K} \delta\vec{m}_k$$

where $\delta\vec{m}_k$ is the change in $\vec{m}_k$, given by

$$\delta\vec{m}_k = \vec{m}_k^{(new)} - \vec{m}_k^{(old)}.$$

We can stop the loop if

$$|\mathbf{\Delta M}| \leq \alpha |\mathbf{M}^{(old)}| \tag{6.2}$$

where α is a small value such as 0.01 or 0.001. Instead of computing the straightforward difference, which may be positive or negative, cancelling out some of the errors, it is a better idea to compute either the absolute value of the error $|\delta\vec{m}_k|$ or the square of the error $|\delta\vec{m}_k|^2$ and compute the sum. Taking the absolute value or the square ensures that the considered error is never negative.

Suppose we perform a total of T iterations of the K-means clustering algorithm given above. In each iteration, we have K means or prototypes. We have to compute the distance between each of the K means and each of the N data points. This takes $O(KN)$ amount of time. Since we have to perform T such iterations, the total amount of computation is $O(KNT)$. If we use the other option for convergence to stop the iterations, it becomes difficult to know the total number of iterations necessary. However, even if the algorithm is looking for convergence, it is a good idea to provide the maximum number of iterations, T, so that the algorithm terminates, no matter what.

6.3   Cluster Quality

It is often necessary to assess the overall quality of a group of clusters produced by a clustering algorithm. When different clustering algorithms are used to cluster a dataset or the same algorithm is used with different parameters, the clusters produced are likely to be different. Different parameters for the same algorithm, and different algorithms are likely to produce different numbers of clusters from the same dataset. Even when different algorithms or different versions of the same algorithm produce the same number of clusters, the composition of the different individual clusters, in terms of actual data points contained in them, is likely to be different. In the case of K-means clustering, the number of clusters produced in one run of the algorithm, i.e., the value of K is known and is provided by the user. However, we can run K-means on the same dataset, producing different numbers of clusters by providing different K values in different runs. We can measure the qualities of the sets of clusters produced in the different runs to pick the set that has the best quality. This approach can be used to determine the most appropriate value of K.

There are two ways to assess the quality of a set of clusters: *intrinsic* and *extrinsic*. Intrinsic assessment computes the quality of a set of clusters by considering the similarities and dissimilarities among the clusters without reference to any external source of knowledge. In particular, an intrinsic quality metric computes a high score for a cluster set for which individual clusters contain data points that are highly similar using a distance or similarity measure, and different clusters are highly dissimilar. A drawback of intrinsic metrics is that although the actual clusters may demonstrate desired quality in terms of a similarity or distance measure, the actual clusters may not be valid as far as domain properties go. For example, the structures or shapes of clusters obtained by the clustering algorithm may not actually exist in the data, depending on the domain of the data and the task at hand. For example, K-means clustering is known to produce a set of convex clusters, and if convex-shaped clusters are not actually present in the dataset, the actual validity may be low even if we obtain high values for intrinsic assessment quality. Having said that, it is still useful to compute intrinsic quality metrics to compare results of clustering, at least as a preliminary or exploratory step. It is necessary to have prior knowledge of the domain to compute extrinsic metrics of cluster quality. For example, if we have a pre-labeled dataset, where data examples have been binned into classes by experts, we can cluster a portion of the labeled dataset (after removing the labels), called a validation or hold-out (sub)set, and use the labels to obtain extrinsic measures of cluster quality.

6.3.1 Intrinsic Cluster Quality Metrics

A number of intrinsic or internal cluster quality metrics have been proposed over the years. We first present a simple statistical measure of intrinsic cluster quality often used with the K-means algorithm. Then we discuss two additional metrics, *Dunn Index* and *Davies Bouldin Index*, used to assess intrinsic cluster quality.

6.3.1.1 Simple Statistical Intrinsic Cluster Quality Metric

This metric obtains a number that explains how much in the variance in the dataset is explained by clustering. It goes though the computation of a few steps to obtain the final metric's value. This approach computes all squared sum of distances, abbreviated as *SS*. In particular, it computes

- *WSS* or *withinSS* or within cluster squared sum of distances,
- *TWSS* or *tot.withinSS* or total within cluster squared sum of distances,
- *BSS* or *betweenSS* or between clusters squared sum of distances, and
- *TSS* or *totss* or total sum of squared distances.

Given a cluster $C_k, k = 1 \cdots K$, *WSS* or *withinSS* computes the squared sum of distances from each point $\vec{x}_i$ to its mean $\vec{m}_k$.

$$WSS(C_k) = \sum_{\substack{i=1 \\ \vec{x}_i \in C_k}}^{|C_k|} d(\vec{x}_i, \vec{m}_k) \tag{6.3}$$

$$= \sum_{\substack{i=1 \\ \vec{x}_i \in C_k}} \sum_{j=1}^{n} (x_{i,j} - m_{k,j})^2 \tag{6.4}$$

Since our goal is to produce dense clusters, a low value of $WSS(C_k)$ is good for any cluster C_k.

$TWSS(\mathcal{C})$ is the sum of $WSS(C_k), k = 1 \cdots K$.

$$TWSS(\mathcal{C}) = \sum_{k=1}^{K} WSS(C_k) \tag{6.5}$$

Since the goal is to obtain a set of clusters where each cluster is compact , a low value of $TWSS(\mathcal{C})$ is desired.

BSS for a set of found clusters computes the sum of distances between all pairs of cluster means.

$$BSS(\mathcal{C}) = \sum_{\substack{i=1\cdots K \\ j=1\cdots K \\ i \neq j}} d(\vec{m}_i, \vec{m}_j) \qquad (6.6)$$

Since our goal is to obtain clusters that are spread out, a large value of $BSS(\mathcal{C})$ is good for a set of clusters $\mathcal{C}$.

Finally, $TSS(\mathcal{C})$ is the sum of $BSS(\mathcal{C})$ and $TWSS(\mathcal{C})$.

The ratio obtained by dividing $BSS(\mathcal{C})$ by $TSS(\mathcal{C})$ gives the total variance in the dataset that is explained by clustering. This number should be high for a good set of clusters.

6.3.1.2 Dunn Index

Dunn Index computes the ratio D between the smallest inter-cluster distance divided by the largest intra-cluster distance. In particular,

$$D(\mathcal{C}) = \frac{min_{\substack{i=1\cdots K \\ j=1\cdots K \\ i \neq j}} d(rep(C_i), rep(C_j))}{max_{k=1\cdots K} \left\{ max_{i=1\cdots |C_k|, j=1\cdots |C_k|, i\neq j} d(\vec{x}_i, \vec{x}_j) \right\}} \qquad (6.7)$$

Here, $\mathcal{C} = \{C_1, \cdots, C_K\}$ is the set of all clusters produced. The numerator computes the minimum inter-cluster distance. We find it by computing the distance between each pair of distinct clusters and then picking the minimum. To compute the distance between two clusters, we can use the mean of all points in the cluster as the representative of the cluster.

$$rep(C_i) = \frac{1}{|C_i|} \sum_{I=1\cdots|C_i|} \qquad (6.8)$$

The representative for a cluster can be obtained in other ways as well, such as the median of the points in the cluster. The denominator first computes the maximum distance between any two points in cluster C_k, and then computes the maximum of all such large inter-example distances in all clusters. Thus, the denominator is the largest intra-cluster distance. Figure 6.1 shows how the numerator and denominator are computed. Figure 6.1 (a) shows the representatives of clusters as the mean point. Figure 6.1(b) shows that to compute the shortest inter-point distance within a cluster, we compute the distance between every pair of points in the cluster and then find the minimum distance. Since the objective in clustering is to obtain clusters with high inter-cluster similarity, and low inter-cluster similarity, a high value of Dunn index is better for a clustering solution compared with another clustering solution that has a lower Dunn index value. This is because for a good set of clusters, even the

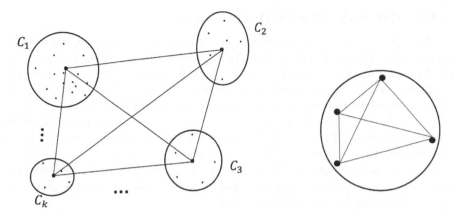

FIGURE 6.1: (a) Computing inter-cluster distance, (b) Computing intra-cluster distance within a single cluster. The intra-cluster distance is computed for all clusters, and the minimum is found.

smallest inter-cluster distance should be as high as possible and the largest intra-cluster distance should be as low as possible, making the ratio of the two, or Dunn index as high as possible. So, for example, to find the "best" value of K for the K-means algorithm, we can try a number of values of K, compute Dunn index for each value, and pick the value of K that gives the largest value of Dunn index.

6.3.1.3 Davies-Bouldin Index

The *Davies-Bouldin Index* is another popular intrinsic cluster metric. Given a set $\mathcal{C} = \{C_1, \cdots, C_k\}$ of clusters produced, it is computed as given below.

$$DB(\mathcal{C}) = \frac{1}{K} \sum_{I=1}^{|\mathcal{C}|} max_{j=1,\cdots|\mathcal{C}|} \left\{ \frac{\sigma_i + \sigma_j}{d(rep(C_i), rep(C_j))} \right\} \qquad (6.9)$$

where σ_i is the average distance from the representative of cluster C_i to the points in cluster C_i. σ_j is the same for cluster C_j. Therefore,

$$\sigma_i = \frac{1}{|C_i|} \sum_{l=1}^{|C_i|} d\left(\vec{x}_{i,l}, rep(C_i) \right) \qquad (6.10)$$

where $|C_i|$ is the number of points in cluster C_i, and $\vec{x}_{i,l}$ is the lth point in cluster C_i.

Since we are looking for a good clustering solution, i.e., a set of clusters produced by an algorithm, such that inter-cluster distances are small and inter-cluster distances are large, the Davies-Bouldin index should be low for a good solution. If we compare two clustering solutions, the one with a lower DB value is better.

6.3.2   Extrinsic Cluster Quality Metrics

Just like for internal assessment, there are several widely accepted metrics for extrinsic evaluation of clustering. As mentioned earlier, extrinsic evaluation of clustering assumes that there is a set of N pre-labeled or pre-classified examples, and the evaluation is performed with respect to this set. The set $\mathbf{D}$ is assumed to contain examples from a number of disjoint classes, $\tilde{C} = \{\tilde{C}_1, \cdots, \tilde{C}_i, \cdots, \tilde{C}_K\}$ into which the data examples are divided. Obviously, when we perform clustering of the same dataset, we do not use the label information. Let the clusters obtained be called $\mathcal{C} = \{C_1, \cdots, C_j \cdots, C_l\}$. The number of clusters found, and the number of actual classes may be different. We discuss two metrics for extrinsic quality assessment for clusters produced by an algorithm: *Purity* and *Rand Index*.

6.3.2.1   Purity

If we look at the data elements in cluster C_j, there may be elements from one or more classes $\tilde{C}_1, \cdots, \tilde{C}_K$. A found cluster C_j may have elements from several classes—all mixed up. However, one class will have the largest number of elements in cluster C_j (if there are ties, we pick the largest class randomly). Let $nmax_j$ be the number of elements in the class with the largest number of elements in found cluster C_j:

$$nmax_j = \underset{\substack{k=1\cdots K \\ x_l \in C_j}}{\arg\max} |\{x_l \in \tilde{C}_k\}| \tag{6.11}$$

where $|\{x_l \in \tilde{C}_k\}|$ is the number of elements in cluster C_j that belong to class $\tilde{C}_k$. $\frac{nmax_j}{|C_j|}$ gives the "purity" of cluster C_j—it is the fraction of elements from the largest contained class in cluster C_j. This is because if we assume that the label of a cluster is that of the most frequent class within it, we can think of $\frac{nmax_j}{|C_j|}$ as the fraction of objects that have been correctly clustered. To compute *purity* of the computed set of clusters, we add the values of $nmax_j$ for each of $j = 1 \cdots l$, and divide by the total number of elements N in the dataset:

$$purity(\mathcal{C}) = \frac{1}{N} \sum_{j=1}^{l} \frac{max_j}{|C_j|} = \frac{1}{N} \sum_{j=1}^{l} \frac{\underset{\substack{x \in 1 \cdots K \\ x_l \in C_j}}{\arg\max} |\{x_l \in \tilde{C}_i\}|}{|C_j|}. \tag{6.12}$$

The highest value that purity can take is 1. This happens if $k = l$ and each found cluster matches exactly one of the pre-labeled classes. Unfortunately, this can also happen if we obtain N clusters with each cluster containing exactly one data example and each data example belonging to a cluster. Thus, a purity of 1 does not guarantee the best clustering possible. As a result, when using purity as a metric, we must take care to see that the number of clusters obtained is as close to the number of actual clusters in the data. In other words, the number of found clusters needs to be kept limited and under control.

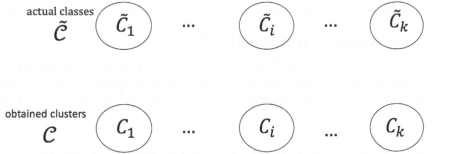

FIGURE 6.2: (a) Actual classes in the dataset based on labels, (b) Found clusters in the dataset, without using the labels.

6.3.2.2 Rand Index

We discuss another extrinsic metric for assessing the quality of a clustering attempt, called Rand Index (RI). To compute the value of Rand Index, we consider all pairs of data elements from the dataset $\mathbf{D}$ containing N elements. Consider a pair or data elements $\langle \vec{x}^{(i)}, \vec{x}^{(j)} \rangle, i, j = 1 \cdots N, i \neq j$. If $\vec{x}^{(i)}$ and $\vec{x}^{(j)}$ belong to the same (labeled) class, and turn out to be in the same cluster, the pair $\langle \vec{x}^{(i)}, \vec{x}^{(j)} \rangle$ is considered a true positive (TP) as far as clustering goes; otherwise, the pair is considered a false positive (FP). Similarly, if $\vec{x}^{(i)}$ and $\vec{x}^{(j)}$ belong to two classes, but turn out to be in two different clusters, the pair $\langle \vec{x}^{(i)}, \vec{x}^{(j)} \rangle$ is considered a true negative (TN); otherwise, the pair is considered a false negative (FN). Rand Index is the fraction of the time we get either a TP or a TN after clustering. Thus,

$$RI = \frac{TP + TN}{TP + TN + FP + FN}. \tag{6.13}$$

If there are N data examples to be clustered, $TP + TN + FP + FN = \binom{N}{2} = \frac{1}{2}N(N-1)$ since we can pick $\binom{N}{2}$ pairs of elements. Therefore,

$$
\begin{aligned}
RI &= \frac{TP + TN}{\frac{1}{2}N(N-1)} \\
&= \frac{2(TP + TN)}{N(N-1)} \\
&\approx \frac{2(TP + TN)}{N^2}
\end{aligned}
\tag{6.14}
$$

6.4   K-Means Clustering in R

There are a number of libraries in R that can perform K-means clustering. In this section, we discuss how several of these libraries can be used to perform K-means clustering, and perform intrinsic and extrinsic assessments of clustering.

6.4.1   Using Extrinsic Clustering Metrics for K-Means

The `Stat` package that comes with the standard distribution provides an implementation of K-means. This library has many statistical and some machine learning functions. Below, we use this library to cluster the Iris dataset after removing its label (*species*) column. We plot the clusters and discuss the quality of the resulting set of clusters. For quality assessment, we use a library called `c2c` that computes extrinsic assessment. In this case, since we have the labels for each of the examples, it is easy to compute purity. The `c2c` library does not compute Rand Index. For plotting, we use the `ggplot` library.

```
1   #file KmeansStatPackage.R
2   #R provides a number of datasets, one of them is the iris dataset
3   library(datasets)
4   head(iris)
5
6   #ggplot2 lets us plot the dataset, plot each of the classes by label Species
7   library(ggplot2)
8   ggplot(iris, aes(Petal.Length, Petal.Width, color = Species)) + geom_point()
9
10  #Perform clustering, we set seed for random number generation for initial clustering
11  set.seed(20)
12  irisCluster <- kmeans(iris[, 3:4], 3, nstart = 20)
13
14  #Provide details about the clusters obtained
15  irisCluster
16
17  #Give a table of clusters and classes. This is the confusion matrix.
18  confusionMatrix <- table(irisCluster$cluster, iris$Species)
19  confusionMatrix
20
21  #Plot the clusters
22  irisCluster$cluster <- as.factor(irisCluster$cluster)
23  ggplot(iris, aes(Petal.Length, Petal.Width, color = irisCluster$cluster)) + geom_point()
24
25  #Use the package called c2c to compute clustering metrics
26  library (c2c)
27
28  #Calculate clustering metrics
29  c2c::calculate_clustering_metrics(confusionMatrix)
30
31  #Compute Rand index using phyclust library
32  library (phyclust)
33  phyclust::RRand (as.integer(iris$Species), as.integer(irisCluster$cluster))
34
35  ############
36  #Cluster the dataset using all features, compute purity and Rand Index
37  irisCluster <- kmeans(iris[, 1:4], 3, nstart = 20)
```

```
38 | irisCluster
39 |
40 | confusionMatrix <- table(irisCluster$cluster, iris$Species)
41 | confusionMatrix
42 | c2c::calculate_clustering_metrics(confusionMatrix)
43 | phyclust::RRand (as.integer(iris$Species), irisCluster$cluster)
```

To refamiliarize, the Iris dataset has 150 examples of three different kinds of iris flowers—setosa, versicolor, and virginica. Each example is described in terms of four features: sepal length, sepal width, petal length and petal width. Each example also has a label associated with it. Thus, this dataset can be used for classification as we have done in previous chapters. We can use it for clustering as well, if we leave out the label *species*. In fact, because the dataset has an associated label, we can use it to perform extrinsic quality evaluation.

The first block of code in lines 1-4 uses a dataset provided within R Studio and print a few rows from the top of the dataset to give a feel for what the dataset looks like.

	Sepal.Length	Sepal.Width	Petal.Length	Petal.Width	Species
1	5.1	3.5	1.4	0.2	setosa
2	4.9	3.0	1.4	0.2	setosa
3	4.7	3.2	1.3	0.2	setosa
4	4.6	3.1	1.5	0.2	setosa
5	5.0	3.6	1.4	0.2	setosa
6	5.4	3.9	1.7	0.4	setosa

Because we cluster this dataset, it is interesting to observe how the data are distributed first. Since the data is 4-dimensional, we randomly choose two features, petal length and petal width, to plot the dataset (see Figure 6.3 (a)). Each species has a different color and each example is plotted as a small circle. We see that setosa is well-separated from the other two classes, but there is some overlap between versicolor and virginica.

K-means clustering starts with a random choice of the cluster means. Therefore, we set a seed to generate a sequence of random numbers in line 11. This ensures that the same random sequence is generated when we run the program several times. Running K-means is simple and is accomplished by line 12, which is repeated here.

```
irisCluster <- kmeans(iris[, 3:4], 3, nstart = 20)
```

To compare results by plotting, we do not cluster the dataset with all its columns, but as specified as the first argument: columns 3 and 4 only—petal length and petal width, respectively. The second argument asks kmeans to find three clusters. We do so to compare the clustering results with the labels from the dataset. The third argument nstart asks the program to run kmeans 20 times and report the best results. The result of clustering is stored in the variable irisCluster. There are several components in the clustering result. Line 15 prints the contents of the irisCluster variable as the following. It prints how many examples belong to each of the clusters.

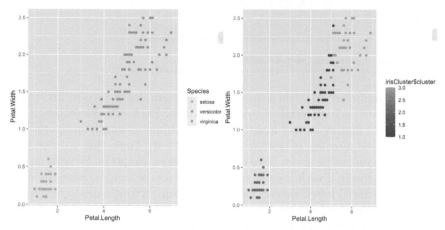

(a) Plot of the Iris Dataset with Two Features   (b) Plot of Clusters with Two Features

FIGURE 6.3: Comparing the Actual Classes and Clusters Produced

In this case, all three clusters are of about the same size. It then prints the means of the three clusters, followed by a vector that shows which example belongs to which cluster. Following this, it prints the within cluster sum of squares and ratio $\frac{BSS}{TSS}$ as described earlier. In this case, 94.3% of the variance in the sum of squares is explained by clustering. At the bottom it says that the `irisCluster` variable contains other components such as the various sum of squares distances, and the number of iterations that `kmeans` went through to produce the clustering.

```
K-means clustering with 3 clusters of sizes 50, 52, 48

Cluster means:
   Petal.Length Petal.Width
1      1.462000    0.246000
2      4.269231    1.342308
3      5.595833    2.037500

Clustering vector:
  [1] 1 1 1 1 1 1 1 1 1 1 1 1 1 1 1 1 1 1 1 1 1 1 1 1 1 1 1 1 1 1 1 1 1 1
 [35] 1 1 1 1 1 1 1 1 1 1 1 1 1 1 1 1 2 2 2 2 2 2 2 2 2 2 2 2 2 2 2 2 2 2
 [69] 2 2 2 2 2 2 2 2 3 2 2 2 2 2 3 2 2 2 2 2 2 2 2 2 2 2 2 2 2 2 2 2 3 3
[103] 3 3 3 3 2 3 3 3 3 3 3 3 3 3 3 3 3 3 2 3 3 3 3 3 3 2 3 3 3 3 3 3 3 3
[137] 3 3 2 3 3 3 3 3 3 3 3 3 3 3

Within cluster sum of squares by cluster:
[1]  2.02200 13.05769 16.29167
 (between_SS / total_SS =  94.3 %)

Available components:

[1] "cluster"     "centers"     "totss"       "withinss"
```

```
[5] "tot.withinss" "betweenss"     "size"          "iter"
[9] "ifault"
```

Line 18 creates a table showing which cluster contains examples from which class.

```
confusionMatrix <- table(irisCluster$cluster, iris$Species)
```

If we print the confusion matrix, we get the table shown below.

```
>   confusionMatrix
    setosa versicolor virginica
1     50          0         0
2      0          2        46
3      0         48         4
```

The confusion matrix says that 50 of the 50 setosas were put in cluster 1. 48 of the 50 versicolors were put in cluster 3, and 2 were put in cluster 2. 46 of the virginicas were placed in cluster 2, with the remaining put in cluster 3. Thus, cluster 1 is pure while cluster 2 and 3 are a little "unpure".

Line 23 plots the actual clusters produced. The cluster plot is in Figure 6.3 (b). We use a library called. c2c to compute clustering metrics. The library's functions can take results from two classification and/or clustering algorithms and compare results. In this case, the class labels in the Iris dataset as well as the cluster memberships are "hard" in that results or labels are given definitively, and not in terms of probabilities of membership in more than one class or cluster, in which case, they are called "soft" memberships. The c2c library can compare results in both cases. Line 29 calculates extrinsic clustering metrics and prints them. The purity metric's values are given below.

```
$overall_purity
[1] 0.96

$class_purity
$class_purity$row_purity
        1         2         3
1.0000000 0.9583333 0.9230769

$class_purity$col_purity
   setosa versicolor   virginica
     1.00       0.96        0.92
```

The overall purity value comes out high at 0.96.

Lines 31-33 compute the Rand Index using the `phyclust` library's `RRand` function.

```
   Rand adjRand  Eindex
 0.9495  0.8857  0.2149
```

We have not discussed adjust Rand Index and E Index in this book.

The clustering above was performed with only two columns. Next, we perform clustering (line 37) with all columns and get the following confusion matrix (line 40).

```
  setosa versicolor virginica
1     0          2        36
2    50          0         0
3     0         48        14
```

This time the clustering results are quite a bit worse than clustering with only two columns. The purity values (line 42) come out worse also.

```
$overall_purity
[1] 0.8933333

$class_purity
$class_purity$row_purity
        1         2         3
0.9473684 1.0000000 0.7741935

$class_purity$col_purity
    setosa versicolor  virginica
      1.00       0.96       0.72
```

This is because the use of all four columns makes the clustering process more difficult since the distance computations are not clear cut like before. The Rand Index also comes out worse.

```
 Rand adjRand  Eindex
0.8797  0.7302  0.2231
```

6.4.2   Using Intrinsic Clustering Metrics for K-Means

In this section, we show how the Nbclust library can be used to compute a number of intrinsic metrics of cluster quality. We have seen the squared sums of distances returned by the kmeans function of the Stat package. In the program below, we show how we can compute additional metrics for intrinsic assessment of clusters, such as Davies-Bouldin and Dunn indices. The Nbclus library can be used to compute a large number intrinsic metrics of clustering and choose a number of clusters based on majority voting.

```
1  #file KmeansNumberofClustersNbclus.R
2  #R provides a number of datasets, one of them is the iris dataset
3  library(datasets)
4  head(iris)
5
6  # Remove species column (5) and scale the data; this is usually needed if
7  #data features are of various magnitudes
8  iris.scaled <- scale(iris[, -5])
```

```
 9
10  #install this package to use a lot of intrinsic ways to determine the
11  #number of clusters
12  library ("NbClust")
13
14  #Find the best partition for k-means using Davies Bouldin index
15  resDB<-NbClust(iris.scaled, diss=NULL, distance = "euclidean", min.nc=2,
16    max.nc=6, method = "kmeans", index = "db")
17
18  #Print results and best number of clusters
19  resDB
20
21  #Find the best partition for k-means using Dunn index
22  resDunn<-NbClust(iris.scaled, diss=NULL, distance = "euclidean", min.nc=2,
23    max.nc=6, method = "kmeans", index ="dunn")
24  resDunn
25
26  #Run 30 different indices to pick the best number of indices
27  resAll<-NbClust(iris.scaled, diss=NULL, distance = "euclidean", min.nc=2,
28    max.nc=6, method = "kmeans", index = "all")
29   resAll
```

Lines 3 and 4 load the Iris dataset. Line 8 scales all the values in the dataset. If the values of the features are of different magnitudes, often scaling them helps clustering as well as classification. Line 12 loads the Nbclust library so that intrinsic metrics of cluster quality can be computed.

Line 15 performs kmeans clustering, varying the number of clusters from 2 to 6. It uses Davies-Bouldin Index to compute the intrinsic quality, prints the Davies-Bouldin Index value for the different numbers of clusters. Line 19 prints the information given below.

```
$All.index
     2      3      4      5      6
0.6828 0.9141 0.9984 1.0526 1.1478

$Best.nc
Number_clusters      Value_Index
       2.0000             0.6828
```

In this case, the best number of clusters is 2, with a Davies-Bouldin Index value of 0.6828.

Lines 22-24 repeats the kmeans clustering process using the Dunn index instead of Davies-Bouldin.

```
$All.index
     2      3      4      5      6
0.2674 0.0265 0.0700 0.0808 0.0842

$Best.nc
Number_clusters      Value_Index
       2.0000             0.2674
```

Dunn index also finds that two clusters are the best.

Lines 27-28 performs intrinsic cluster quality assessment using a number of metrics, most of which we have not discussed in this book. It concludes that the best number of clusters by majority voting of the metrics.

```
*********************************************************************
* Among all indices:
* 9 proposed 2 as the best number of clusters
* 13 proposed 3 as the best number of clusters
* 1 proposed 5 as the best number of clusters
* 1 proposed 6 as the best number of clusters

              ***** Conclusion *****

* According to the majority rule, the best number of clusters is  3

*********************************************************************
```

Additional details for each of the metrics used can be obtained by typing resAll in the command-line prompt.

6.5   Hierarchical or Connectivity-Based Clustering

In partitional clustering such as K-means clustering, the clustering process is iterative, but it ultimately leads to a number ($= K$) clusters, where each cluster is considered to be at the same level. In contrast, in hierarchical clustering, the clustering process starts at one of two extremes: 1) all points are in the same cluster, or 2) each point is a cluster by itself, and then 1) either divide a big cluster into small clusters repeatedly, or 2) coalesce small clusters into bigger clusters repeatedly.

The first case is called *divisive clustering*, and the second *agglomerative clustering*. In either case, the objective is not to have really big or really small clusters, but clusters of "reasonable" sizes. In other words, the division process or the agglomeration process stops after a while based on some criterion. In any case, to divide a big cluster further or to combine a pair of small clusters into a bigger one, we have to compute the distance between two clusters. A cluster usually contains multiple points and computing the distance between two clusters requires computing the distance between two subsets of data points.

Assume that Figure 6.4 shows two clusters during the process of hierarchical clustering—whether divisive or agglomerative—with the distribution of data points, assuming a 2-D situation where each data point has two features x_1 and x_2. Depending on how distance is computed between two multi-point clusters, hierarchical clustering can be characterized in three ways as shown in Figure 6.5.

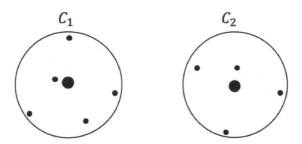

FIGURE 6.4: Distance between two clusters

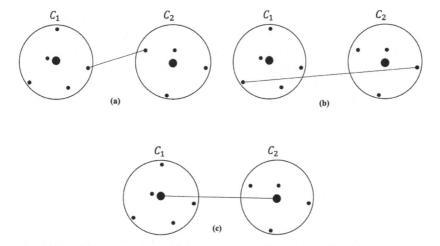

FIGURE 6.5: Three ways to computer distance between two clusters: (a) Using a closest pair of points in the clusters, (b) Using a farthest pair of points in the clusters, and (c) Using the centroids or representatives of the two clusters.

How the distance is computed between two clusters is called the *linkage criterion*. When the distance between two clusters C_1 and C_2 is computed using a closest pair of points, it is called *minimum-linkage* or *single-linkage clustering*. In such a case, the distance d' between C_1 and C_2 is computed as

$$d'(C_1, C_2) = min\left\{d(\vec{x}_1, \vec{x}_2), \forall \vec{x}_1 \in C_1, \vec{x}_2 \in C_2\right\} \qquad (6.15)$$

where d is a metric used to compute the distance between the two points—usually Euclidean distance, and $\vec{x}_1$ and $\vec{x}_2$ are two data points.

If a pair of farthest points between the two clusters is used to compute the distance between two centers, it is called *maximum linkage or complete-linkage clustering*. In such a case, the distance d' between clusters C_1 and C_2

is computed as

$$d'(C_1, C_2) = max\{d(\vec{x}_1, \vec{x}_2), \forall \vec{x}_1 \in C_1, \vec{x}_2 \in C_2\}. \qquad (6.16)$$

If we compute the average distance between points in the two clusters, it is called *average-linkage clustering*:

$$d'(C_1, C_2) = \frac{\sum_{\substack{\vec{x}_1 \in C_1 \\ \vec{x}_2 \in C_2}} d(\vec{x}_1, \vec{x}_2)}{|C_1|\,|C_2|}. \qquad (6.17)$$

If we compute the distance between the two centroids, it is called *centroid-linkage clustering*.

Whether it is divisive or agglomerative hierarchical clustering, the distance computation is used to move the process forward. In agglomerative clustering, each point is initially considered a cluster. Thus, there are N singleton clusters to begin with. The clustering process examines pairwise distance among the N clusters, and picks a pair with a smallest distance between them and merges them. Thus, at this point, we have $N-1$ clusters, of which one has two points and rest are singleton clusters. In the next iteration, distance is computed pairwise among the $N-2$ clusters using a linkage approach, and a pair with the smallest distance is merged to bring down the number of clusters to $N-3$. The process can be continued till we have one single cluster containing all the points.

In either agglomerative or divisive clustering, we can optionally cut off the process at a certain point. Figure 6.6 shows the relative distances being computed among the clusters at various levels as vertical distances. For example, if we perform agglomerative clustering and cut off the tree being formed (it is called a *dendrogram* if relative distances are used to create lengths of tree branches) at the dashed line labeled cut_1, the set of clusters obtained are $\{\{\vec{x}_1, \vec{x}_2\}, \{\vec{x}_3\}, \{\vec{x}_4, \vec{x}_5\}\}$. The distance information may or may not be used, depending on the domain. For example, if we cluster a number of students at a university using various characteristics, the relative distances in the dendrogram may not matter. However, if we are clustering a number of living species in the world, the distances may correspond to evolutionary distances among species. Divisive clustering, if cut off at the same cut point, we will have two clusters: $\{\{\vec{x}_1, \vec{x}_2\}, \{\vec{x}_3\,\vec{x}_4, \vec{x}_5\}\}$.

There is a choice of approaches to compute the distance between two data points, the most common being Euclidean distance as mentioned earlier. There are also choices for how to compute the distance between a pair of clusters as discussed earlier. In agglomerative clustering, since the distance between a pair of clusters is computed, and to start with have N clusters, $\frac{1}{2}N(N+1)$ or $O(N^2)$ distance calculations are performed. Each subsequent iteration also requires distance calculations between all pairs of currently present clusters, requiring time proportional to the square of the current number of clusters. As the size of a cluster goes up, starting from just 1 to start, the algorithm needs to compute pairwise distance where one element belongs to one cluster and the

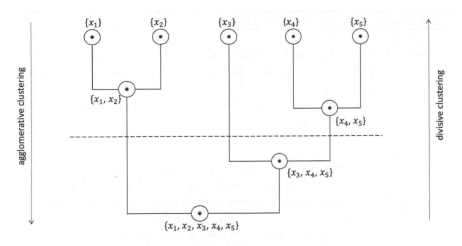

FIGURE 6.6: Divisive and Agglomerative Hierarchical Clustering

---

Algorithm 6.2: Algorithm for Agglomerative Clustering

Input: A dataset of unlabeled points $\mathbf{D} = \{\vec{x}_1, \cdots, \vec{x}_i, \cdots, \vec{x}_N\}$
Output: A set of subsets or clusters $\mathcal{C} = \{C_1, \cdots, C_k\}$ such that
$\sum_{I=1}^{K} = N$ with $\forall \vec{x}_i \in \mathbf{D}, and \vec{x}_i \in C_k$ for some $k = 1 \cdots K$,
and $\forall I, j\ C_i \cap C_j = \phi, I \neq j$

1 //Put each item in a cluster by itself; Name the clusters $C_1 \cdots C_N$
2 $\mathcal{C} \leftarrow \{C_1 = \{\vec{x}_1\}, \cdots, C_N = \{\vec{x}_N\}\}$
3 $nclusters \leftarrow N$
4 **repeat**
5   //Find two clusters that are closest using distance computation d'
    and a linkage approach L. d' is different from d–the metric used
    to compute distance between two points
6      $\langle C_i, C_j \rangle \leftarrow argmin_{\langle C_m, C_p \rangle,\ m \neq p}\ d'(C_m, C_p, L)$
7   //Remove the two clusters C_i and C_j
8      $\mathcal{C} \leftarrow \mathcal{C} - \{C_i, C_j\}$
9   //Add a new cluster containing all the elements of C_i and C_j.
    Arbitrarily, call the new cluster C_i
10     $\mathcal{C} \leftarrow \mathcal{C} \cup \{C_i = \{C_i \cup C_j\}\}$
11   //Decrease the number of clusters
12     $nclusters \leftarrow nclusters - 1$
13 **until** $nclusters == K$;
14 **return** $\mathcal{C}$

Algorithm 6.3: Algorithm for Divisive Clustering

Input: A dataset of unlabeled points $\mathbf{D} = \{\vec{x}_1, \cdots, \vec{x}_i, \cdots, \vec{x}_N\}$

Output: A set of subsets or clusters $\mathcal{C} = \{C_1, \cdots, C_k\}$ such that $\sum_{I=1}^{K} = N$ with $\forall \vec{x}_i \in \mathbf{D}, and \vec{x}_i \in C_k$ for some $k = 1 \cdots K$, and $\forall I, j \; C_i \cap C_j = \phi, I \neq j$

1 //Put all items in a single cluster by itself; Name the cluster C_1

2 $\mathcal{C} \leftarrow \{C_1 = \{\vec{x}_1 \cdots, \vec{x}_N\}\}$

3 $nclusters \leftarrow 1$

4 **repeat**

5   //Find a non-singleton cluster C_j and take out an element $\vec{x}_i$ out of it so that the distance d' between $C_j - \{\vec{x}_i\}$ and $\{\vec{x}_i\}$ is the highest, considering all clusters and all points within the clusters.

6     $C_j, \vec{x}_i = argmax_{i \in C_j, j \in 1 \cdots K} \; d'(C_j - \{\vec{x}_i\}, \{\vec{x}_i\}, L)$

7     $C_{ncluster+1} \leftarrow \{\vec{x}_i\}$ //make a new singleton cluster

8     $C_j \leftarrow C_j - \{\vec{x}_i\}$ //remove the element from cluster C_j

9     $\mathcal{C} \leftarrow \mathcal{C} + C_{ncluster+1}$ //add the new singleton cluster to the cluster set

10 **until** $nclusters == K$;

11 **return** $\mathcal{C}$

---

other belongs to the other cluster. Thus, precise computation of time becomes complex. Ignoring the second set of computations, and focusing on the pairwise distance computation between clusters, the total number of computations is

$$O(N^2) + O((N-1)^2) + \cdots + O((K+1)^2). \tag{6.18}$$

Assuming N to be much larger than K, i.e., we have a fairly large dataset (say with $N = 100,000$), and K is small (e.g., assume $K = 10$), we see that each of the items in the sum above $\approx O(N^2)$, and there are approximately N such items to be summed. This gives the sum as $O(N^3)$. Thus, agglomerative hierarchical clustering, as discussed above, is expensive in terms of time. As a result, agglomerative clustering is not practical even for a moderately small number of data examples. In such a case, K-means clustering becomes more attractive.

    In Algorithm 6.3, the divisive clustering produces K clusters and stops. However, in some cases such as obtaining the evolutionary steps and clustering of species or life forms, the clustering process goes all the way down to one cluster, signifying the initial point in the origin of life. In contrast to an agglomerative algorithm, the divisive algorithm starts with one cluster $\mathcal{C} = \{\{\vec{x}_1, \cdots, \vec{x}_N\}\}$, containing all the data points. The divisive algorithm will form two clusters by taking one data element $\vec{x}_i$ out of the cluster to produce two clusters $\mathcal{C} = \{\{\vec{x}_i\}, \cdots, \{\vec{x}_1, \cdots \vec{x}_{i-1}, \vec{x}_{i+1}, \cdots \vec{x}_N\}\}$. How to choose $\vec{x}_i$ is the question. One possibility is to find $\vec{x}_i$ such that the distance between the two clusters is the largest. This process is repeated as many times as necessary.

In this algorithm, one item at a time is taken out of one of the existing non-singleton clusters. This is a simplest algorithm when once an item $\vec{x}_j$ is taken out of a non-singleton cluster to form a new singleton cluster, it $(\vec{x}_j)$ is never put back in another cluster in the cluster set.

Simple divisive algorithms such as the ones given here can be improved in many ways. For example, it is possible to remove more than one item from an existing cluster to form a new cluster or a singleton cluster may be merged with another existing cluster. The possibilities are endless, but the algorithms become more complex as well.

6.5.1 Agglomerative Clustering in R

We use agglomerative clustering implementation called `hclust` available in the `stats` library in R. The `stats` package comes with R and does not have to be installed separately. In the program given below, we first scale the Iris dataset. In the first part of the program, we randomly pick 20 elements from the dataset and cluster them. We pick a small number of elements to show the dendrogram, which is difficult to show with 150 elements. In the second part of the program, we cluster the entire Iris dataset, compute an extrinsic assessment metric, and plot the clusters.

```
1  #file AgglomerativeClusteringIris1.R
2  #R provides a number of datasets, one of them is the iris dataset
3  library(datasets)
4  head(iris)
5
6  #Remove species column (5) and scale the data; this is usually needed if
7  #data features are of various magnitudes
8  irisScaled <- scale(iris[, -5])
9
10 ##########Cluster a reduced dataset and draw dendogram########
11 #Sample 20 rows from the Iris dataset
12 irisReduced <- sample (irisScaled, 20)
13 #Obtain the dissimilarity matrix for the  reduced dataset
14 dIrisReduced <- dist (irisReduced, method = "euclidean")
15 #Obtain the hierarchical clusters all the way to single elements
16 hIrisReducedClusters <- hclust (dIrisReduced, method = "complete")
17 #Plot the clusters from the reduced dataset
18 plot (hIrisReducedClusters, cex = 0.6)
19
20
21 ##########Cluster the entire dataset####
22 #obtain the dissimilarity matrix
23 dIris <- dist (irisScaled, method = "euclidean")
24 hIrisClusters <- hclust (dIris, method = "complete")
25 #Plot the dendogram, it comes out quite crowded since the number of points is
26 #large for plotting
27 plot (hIrisClusters, cex = 0.6)
28
29 #Cut into three clusters since we do not want to go down to singleton clusters
30 firstClusters <- cutree(hIrisClusters, k=3)
31 #Cut at a height of 3 instead
32 secondClusters <- cutree(hIrisClusters, h=3)
33
34 #Show how many points are in each cluster
```

```
35  table (firstClusters)
36  table (secondClusters)
37
38  #Create a confusion matrix for the first cluster set
39  confusionMatrix <- table(firstClusters, iris$Species)
40  confusionMatrix1
41
42  #Compute purity, extrinsic clustering metric for the first cluster set
43  library (c2c)
44  c2c::calculate_clustering_metrics(confusionMatrix)
45
46  #Use factoextra library for plotting clusters
47  library (factoextra)
48
49  #Plot the first set of clusters
50  fviz_cluster(list(data=dIris, cluster = firstClusters))
```

Lines 3 and 4 load the Iris dataset, available directly within R. Line 8 removes the class label in the dataset and scales it so that all the dimensions are comparable. Lines 10-18 extract 20 random examples from the scaled unlabeled dataset (line 12), obtain a dissimilarity or distance matrix using Euclidean distance (line 14), obtain a hierarchical clustering of the reduced dataset using the complete linkage approach (line 16), and plots the clusters obtained from the dataset. The distance matrix stores the Euclidean distance between any two examples in the dataset. Clustering is performed based on this distance matrix. Agglomerative clustering starts with singleton clusters (clusters containing single elements) and then coalesces them one at a time, to build a single cluster out of all the elements. This is shown in the plot of the clustering process, represented as a tree called dendrogram. The dendrogram for the reduced Iris dataset is shown in Figure 6.7 (a). Since the dataset has

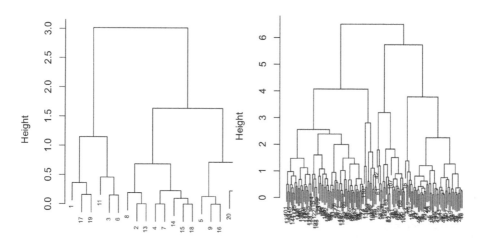

(a) Dendrogram of the Reduced Iris Dataset   (b) Dendrogram of the Full Iris Dataset

FIGURE 6.7: Dendograms for the Iris Dataset

only 20 examples, the dendrogram can be fully drawn on a piece of paper. The dendrogram starts with singleton clusters at the bottom and progressively adds one element to a cluster at a time, to create a single cluster on top containing all the elements. The vertical direction shows the distance between the clusters.

The reduced dataset was created so that the corresponding dendrogram tree can be easily shown. The dendogram is important in some cases such as the study of evolutionary relationship among species based on genetic. In such cases, the evolutionary distance, which shows how much time (in some unit) may have passed for speciation, may be important in some cases. However, the dendrogram may not play a significant role in other situations and need not be drawn, where discovering the clusters is important. Lines 21-27 cluster the entire Iris dataset using hierarchical clustering, and plot the corresponding dendrogram (see Figure 6.7 (b)), which is dense and difficult to see due to the number of examples involved.

Lines 29-36 show how to cut the dendrogram in a couple of different ways. Line 30 cuts the dendrogram in such a way as to obtain a set of three clusters called `firstClusters` using the function called `cutter`. Here k is like the number of clusters in K-means clustering. Line 31 cuts the tree at a height of 3 to form a set of clusters called `secondClusters`. In lines 35 and 36, the program shows how many points are in each of the two cluster sets produced. In this case, `firstClusters` contains three clusters of size 49, 24 and 77, and `secondClusters` contains six clusters of size 42, 7, 5, 66, 19 and 11.

```
> table (firstClusters)
firstClusters
 1  2  3
49 24 77
> table (secondClusters)
secondClusters
 1  2  3  4  5  6
42  7  5 66 19 11
```

Lines 38-44 create a confusion matrix between the clustering results from the first set of clusters and the actual labels in the Iris dataset, and compute purity.

Lines 47-50 use an R library called factoextra to print the examples first set of clusters. The clusters are shown in Figure 6.8. The two dimensions used are the two principal components of the dataset. Although we have not discussed how principal components can be computed in this book, in this case, the first dimension explains 71.5% of the variance in the dataset and the second dimension explains another 20%. Thus, between the two dimensions, 91.5% of the variance is explained.

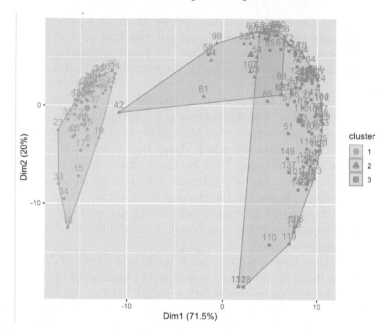

FIGURE 6.8: Agglomerative Hierarchical Clusters of the Iris Dataset

6.5.2   Divisive Clustering in R

Divisive hierarchical clustering is implemented in a few libraries in R. One such implementation is in the `cluster` library in terms of an algorithm called DIANA (Divisive Analysis) of Clustering. The algorithm is similar to the divisive algorithm discussed earlier. In the program below, DIANA is used to cluster the Iris dataset after removing the labels. It requires a distance or dissimilarity matrix (line 14) like the implementation of the agglomerative algorithm discussed in the previous section. Clustering is performed in line 15. The dendrogram can be plotted easily (line 18), and clusters can be obtained by cutting the dendrogram , either using a value for k—the number of clusters desired, h—the height at which the dendrogram is cut. Using the `c2c` library, purity can be computed (line 35) from a confusion matrix showing how the clustering results can be compared with class labels in the original Iris dataset. Using the `factoextra` library, the clusters can be printed (line 41).

```
1   #file DivisiveClusteringIris.R
2   #R provides a number of datasets, one of them is the iris dataset
3   library(datasets)
4   head(iris)
5
6   library (cluster) #implements Diana and a number of other clustering algorithms
7
8   #Remove species column (5) and scale the data; this is usually needed if
9   #data features are of various magnitudes
10  irisScaled <- scale(iris[, -5])
11
```

```
12  ##########Cluster the entire dataset####
13  #obtain the dissimilarity matrix
14  dIris <- dist (irisScaled, method = "euclidean")
15  divisiveIrisClusters <- diana (dIris, metric = "euclidean")
16  #Plot the dendogram, it comes out quite crowded since the number of points is
17  #large for plotting
18  plot (divisiveIrisClusters)
19
20  #Cut into three clusters since we do not want to go down to singleton clusters
21  firstClusters <- cutree(as.hclust(divisiveIrisClusters), k=3)
22  #Cut at a height of 3 instead
23  secondClusters <- cutree(as.hclust(divisiveIrisClusters), h=3)
24
25  #Show how many points are in each cluster
26  table (firstClusters)
27  table (secondClusters)
28
29  #Create a confusion matrix for the first cluster set
30  confusionMatrix <- table(firstClusters, iris$Species)
31  confusionMatrix
32
33  #Compute purity, extrinsic clustering metric for the first cluster set
34  library (c2c)
35  c2c::calculate_clustering_metrics (confusionMatrix)
36
37  #Use factoextra library for plotting clusters
38  library (factoextra)
39
40  #Plot the first set of clusters
41  factoextra::fviz_cluster(list(data=dIris, cluster = firstClusters))
```

6.6 Density-Based Clustering

The K-means clustering algorithm, discussed earlier, can find only spherical clusters. The hierarchical algorithms are also likely to find spherical clusters only if we use a distance metric like Euclidean distance. However, clusters can be of arbitrary shapes in some domains or tasks. Figure 6.9 shows some clusters of different shapes. The K-means algorithm or a hierarchical algorithm will not be able to find such clusters in a dataset. These algorithms are also messed up by the presence of outliers. This is where other types of algorithms such as density-based clustering come to help. Density-based clustering looks for regions of high density within the dataset. The hypothesis is that clusters in a dataset occur in areas of high density. A dataset, when visualized, is likely to have areas of high density, possibly of arbitrary shapes, interspersed with low-density or sparse regions. The high-density regions are likely to contain identifiable clusters whereas low-density regions or sparse regions may contain outliers that do not belong to any clusters. Of course, not all clusters have to be of the same density. However, all clusters have to have a minimum level of density. DBSCAN is the pioneering algorithm in density-based clustering and we discuss it below.

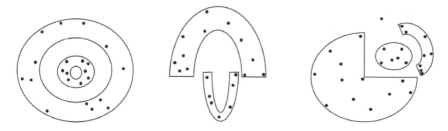

FIGURE 6.9: Non-convex Cluster Shapes

6.6.1 DBSCAN

Usually, we understand density. To be a number of points or items per unit area or space. For example, we can compute the population density of a city, state or country per square mile or kilometer. DBSCAN defines density not for a region with the distribution of data points, but defines density for each point in the dataset. DBSCAN uses two user-provided parameters called *nPoints* and $\epsilon > 0$ to compute density. ϵ provides the radius of a neighborhood around a data point, and is computed using a metric like Euclidean distance. The number of points within ϵ radius of a point $\vec{x}_i$ gives the density of the point, as shown in Figure . The algorithm computes the density of all N points in the dataset. It also keeps a record of which points are in the ϵ-sized neighborhood (or, simply ϵ-neighborhoold) of $\vec{x}_o$.

If the number of points in the ϵ-neighborhood of a point $\vec{x}_i \geq nPoints$, $\vec{x}_i$ is called a *core point* within the dataset.

$$|\epsilon\text{-}neighborhood(\vec{x}_i)| \geq nPoints \implies core\text{-}point(\vec{x}_i) \qquad (6.19)$$

Thus, given a dataset, and user-provided values of *nPoints* and ϵ, we can make a pass over each of the points and determine whether it is a core point, using a distance metric. In density-based clustering, clusters are formed around core points; hence, the name.

The idea of *reachability* is also central to density-based clustering. There are two kinds of reachability that DBSCAN defines: i) direct-reachability and ii) density-reachability. Given a point $\vec{x}_i$, all points $\vec{x}_j$ within the ϵ-neighborhood are defined to be *directly-reachable* from $\vec{x}_i$. We can denote this as $\vec{x}_i \rightarrow \vec{x}_j$.

Given point $\vec{x}_j$ that is directly-reachable from $\vec{x}_i$, we can construct an ϵ-neighborhood around $\vec{x}_j$ as well. Any point $\vec{x}_l$ in the ϵ-neighborhood of $\vec{x}_j$ is also defined as *density-reachable* from $\vec{x}_i$. We can denote this as $\vec{x}_i \rightsquigarrow \vec{x}_i$. We can repeat the process of computing ϵ-neighborhoods. Any point $\vec{x}_l$ that can be reached from $\vec{x}_i$ by a chain of direct-reachability relations is *density-reachable* from $\vec{x}_i$. This is described in the logical formulas given

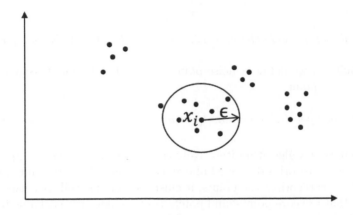

FIGURE 6.10: ϵ-neighborhood of a point $\vec{x}_i$

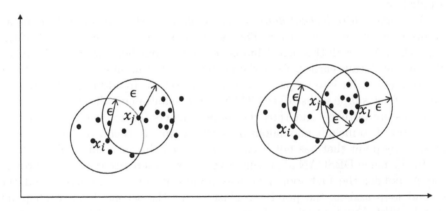

FIGURE 6.11: Direct-reachability as defined by DBSCAN

in Equation 6.20.

$$\forall \vec{x}_i, \vec{x}_j \;\; \vec{x}_j \in \epsilon\text{-}neighborhood(\vec{x}_i) \implies directly\text{-}reachable(\vec{x}_i, \vec{x}_j) \tag{6.20}$$

$$\forall \vec{x}_i, \vec{x}_j \;\; directly\text{-}reachable(\vec{x}_i, \vec{x}_j) \implies density\text{-}reachable(\vec{x}_i, \vec{x}_j)$$

$$\forall \vec{x}_i, \vec{x}_j, \vec{x}_l \;\; density\text{-}reachable(\vec{x}_i, \vec{x}_j) \wedge$$
$$directly\text{-}reachable(\vec{x}_j, \vec{x}_l) \implies density\text{-}reachable(\vec{x}_j, \vec{x}_l)$$

Having defined core points, direct-reachability and density-reachability, we can define the concept of border points. A *border point* $\vec{x}_b$ is a point that is density-reachable from a point $\vec{x}_i$, but $\vec{x}_b$ does not have the density to be a core point.

$$\forall \vec{x}_i, \vec{x}_b \ \ density\text{-}reachable(\vec{x}_i, \vec{x}_b) \land \neg core\text{-}point(\vec{x}_b) \implies border\text{-}point(\vec{x}_b)$$
$$(6.21)$$

In addition, we define a noise-point as one that is not density-reachable from any core point.

$$\forall \vec{x}_n \ \left\{ \forall \vec{x}_c \ core\text{-}point(\vec{x}_c) \land \neg density\text{-}reachable(\vec{x}_c, \vec{c}_n) \right\} \implies noise\text{-}point(\vec{x}_n)$$
$$(6.22)$$

The DBSCAN algorithm iteratively categorizes each point $\vec{x}_i$ in a dataset **D** into three categories discussed above: core points, border points and noise points. As it categorizes the points, it clusters them as well. A cluster is a set of points that are core points, and points that are density-reachable from core points—including more core points and border points, such that the points in the cluster do not belong to any other clusters. This ensures that the clusters are disjoint.

The function *collect-all-density-reachable-points*($\vec{x}_i$) recursively finds all points that belong to a cluster. This is the most time-consuming part of the algorithm. Although this algorithm is not spelled out here, we have discussed all relevant details earlier. As it should be clear the number of clusters to be found does not have to be pre-specified in DBSCAN. In addition, the clusters can be of any shape and size. DBSCAN also can automatically discover points it thinks are outliers, and thus the found clusters do not get misshaped by being forced to include anomalous or outlier points in them. In the algorithm above, any point that has not been assigned to a cluster is an outlier point.

In the naive DBSCAN algorithm given above, there are two loops through the N points: the first loop identifies points as core or non-core, and the second loop assigns the points to clusters. In the second loop, each point may be visited multiple times. However, the most amount of time is spent on finding the points within an ϵ-neighborhood of a point. This is often called a *range query*. Careful analysis shows that the neighborhood or range query computation needs to be performed only once for each data point in the entire algorithm. To perform this in a straightforward manner, the algorithm will have to pre-compute the distances between each pair of points and keep the distances in an $N \times N$ matrix for N points. This takes $O(N^2)$ time and $O(N^2)$ memory as well. Given such a matrix, the algorithm can find points lying within an ϵ-neighborhood of a point in $O(N)$ time by scanning over all other points. Thus, a chain of distance computations needed to find points in a cluster can be done efficiently. The overall time complexity and memory (space) complexity has been shown to be $O(N^2)$ in such an implementation. Since the algorithm requires finding if a point is within a certain distance of another point, the use of an efficient data structure makes the implementation much more efficient. A data structure like an R*-tree can be used to find ϵ-neighbors of a point in $O(N \ logN)$ time. Efficient implementation of DBSCAN have been reported to take $O(N \ log \ N)$ and $O(N)$ memory.

Algorithm 6.4: Naive Algorithm for Density-based Clustering

Input: A dataset of unlabeled points $\mathbf{D} = \{\vec{x}_1, \cdots, \vec{x}_i, \cdots, \vec{x}_N\}$
Output: A set of subsets or clusters $\mathcal{C} = \{C_1, \cdots, C_k\}$ such that
$\sum_{I=1}^{K} = N$ with $\forall \vec{x}_i \in \mathbf{D}, and \vec{x}_i \in C_k$ for some $k = 1 \cdots K$,
and $\forall I, j \; C_i \cap C_j = \phi, I \neq j$

1 //Identify each point as a core point or non-core point
2 **for** $i = 1$ **to** N **do**
3   //no point has been assigned to a cluster yet
4   //keep a vector saying which cluster it is assigned to, -1 means
    not assigned yet $assigned\text{-}to\text{-}cluster\text{-}p_i \leftarrow -1$
5   //Keep a vector indicating if a point is core or not
6   **if** $|\epsilon\text{-}neighborhood| \geq nPoints$ **then**
7 | $core\text{-}point\text{-}p_i \leftarrow true$
8   **else**
9      $core\text{-}point\text{-}p_i \leftarrow false$

10 $nclusters \leftarrow 0$ //no clusters yet
11 //Go over each point again
12 **for** $i = 1$ **to** N **do**
13   **if** $assigned\text{-}to\text{-}cluster\text{-}p_i$ **then**
14 | skip point $\vec{x}_i$ //Already assigned to a cluster
15   **else if** $core\text{-}point\text{-}p_i$ **then**
16      $ncluster \leftarrow ncluster + 1$
17      $clusterPointSubset \leftarrow collect\text{-}all\text{-}density\text{-}reachable\text{-}points(\vec{x}_i)$
18   **foreach** $\vec{x}_j \in lcusterPointSubset$ **do**
19        $assigned\text{-}to\text{-}cluster\text{-}p_i \leftarrow ncluster$

20 **return** $\mathcal{C}$

6.6.2 Density-Based Clustering in R

The following program clusters the Iris dataset after removing the `Species` column. It does so using the `dbscan` function in the `dbscan` library. DBSCAN is very finicky about the values of the two parameters: `eps` and `minPts`. The clustering results can vary dramatically depending on the two values. There is published literature on how one may go about finding appropriate values for the two parameters' values. In this program, we simply perform a grid search to do so, and use the "best" parameters found, considering the overall purity of the set of clusters, to cluster the dataset.

```
1  #file DensityClusteringIris.R
2  #R provides a number of datasets, one of them is the iris dataset
3  library(datasets)
4  head(iris)
5
6  library (dbscan) #implements a fast version of DBSCAN
7
```

```
 8 #Remove species column (5) and scale the data; this is usually needed if
 9 #data features are of various magnitudes
10 irisScaled <- scale(iris[, -5])
11
12 ############
13 # Find suitable DBSCAN parameters: Perform grid search
14 library (c2c)
15 epsVals <- seq (0.1, 1, 0.05)
16 minPtsVals <- seq (4, 10)
17 maxPurity <-  0
18 #Loop over a number of values of eps and minPts, and keep track of
19 #the combination that gives maximum value of overall_purity using
20 #c2c library
21 for (epsVal in epsVals) {
22   for (minPtsVal in minPtsVals)
23     {
24         dbscanIrisClusters <- dbscan::dbscan (irisScaled,
25            eps = epsVal, minPts = minPtsVal )
26        confusionMatrix <- table(dbscanIrisClusters$cluster, iris$Species)
27        extrinsicAssess <- c2c::calculate_clustering_metrics(confusionMatrix)
28        purity <- extrinsicAssess$overall_purity
29         if (purity > maxPurity){
30           maxPurity <- purity
31           maxEps <- epsVal
32           maxMinPts <- minPtsVal
33           }
34     }
35   }
36 ###########
37
38 #print values of eps and minPts that give maximum purity
39 print ( c("eps = ",  maxEps, " minPts = ", maxMinPts, " Purity = ", maxPurity))
40
41 #run dbscan with the "best" parameters; we could saved results in the loop
42 dbscanIrisClusters <- dbscan::dbscan (irisScaled, eps = maxEps,
43        minPts = maxMinPts)
44
45 #Plot the  "best" set of clusters
46 library (factoextra)
47 factoextra::fviz_cluster(dbscanIrisClusters, data = irisScaled,
48             geom = "point", ellipse = FALSE)
```

Lines 1-6 load the Iris dataset and scales it after removing the Species column. Lines 12-36 use the c2c library to run through a combination of eps values ranging from 0.1 to 1 in jumps of 0.5, and mints values ranging from 4 to 10. The program starts with a value of 4 because the original authors of DBSCAN recommend it to be the minimum value to try. For each pair of eps and minPts values, the program runs dbscan (lines 24-25), creates a confusion matrix (line 27), and computes overall purity using the c2c library. It keeps track of the eps and minPts values for the combination that gives the highest purity, an extrinsic assessment metric, discussed earlier. It prints the values in line 39. It runs dbscan again for this combination of the two parameters (lines 41-43). Finally, using the factoextra package's fviz_cluster function, it plots the clusters obtained. In this case, the values of eps and minPts are 0.45 and 5 respectively, for optimal purity results.

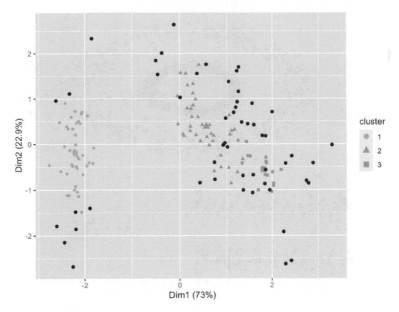

FIGURE 6.12: Clusters Obtained Using DBSCAN in the Iris Dataset

```
DBSCAN clustering for 150 objects.
Parameters: eps = 0.45, minPts = 5
The clustering contains 3 cluster(s) and 54 noise points.

 0  1  2  3
54 41 40 15
```

A plot of the clusters in the Iris dataset obtained using these parameter values is shown in Figure 6.12.

6.6.3 Clustering a Dataset of Multiple Shapes

The Iris dataset is simple in that the present groups are separable fairly well using convex shaped clusters. Centroid-based algorithms like K-means, hierarchical clustering and density-based clustering algorithms are able to separate the clusters reasonably well. However, sometimes the "natural" groups are of irregular shapes that centroid-based and hierarchical clustering algorithms find it impossible to discover. The *multishapes* dataset (Figure 6.13) has been especially synthesized to test clustering algorithms. This dataset has two ellipsoidal clusters on top, one inside the other; two linear clusters in the bottom left; and a small dense cluster in the bottom right; and a number of outlier points. The set of clusters obtained using the K-means algorithm is shown in Figure 6.14. As expected, K-means clustering cannot find the irregular shapes present although it finds the three lower clusters. In place of

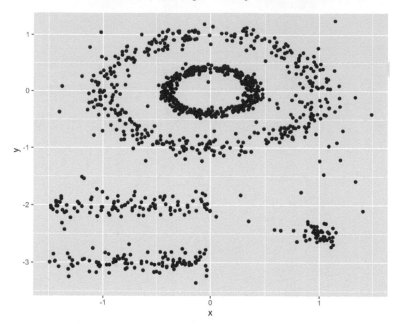

FIGURE 6.13: Plot of the Multishapes Dataset

the two ellipsoidal clusters, it finds three clusters—two on the left side each encompassing parts of the ellipses, and one on the right hand side covering parts of both clusters. This is because K-means clustering looks for spherical clusters in the datas, whether they are present or not.

A program to find clusters using both K-means and DBSCAN in this dataset is given below. The *multishapes* dataset is available for download in R as a part of the `factoextra` package. The dataset has three columns. The first two columns give X and Y values for the points, and the third column gives the shape to which a point belongs.

```
1  #file DensityClusteringMultishapes.R
2  library (dbscan) #implements a fast version of DBSCAN
3
4  #multishapes simulated data from factoextra package
5  library (factoextra)
6  multishapesData <- multishapes[,1:2] #the first two columns give the points
7  multishapesLabels <- multishapes [,3] #the last column gives the labels
8
9  #plot the multishapes data, has two ellipsoidal clusters, one inside the other;
10 #two linear clusters; another small cluster; and a bunch of outliers
11 ggplot (multishapesData, aes (x,y))+geom_point ()
12
13 #K-means clustering of multishapes dataset
14 #Perform clustering, kmeans from the stats package sets initial clusters
15 #randomly and that's why we set seed for random number generation
16 set.seed(20)
17 kmeansMultishapesClusters <- kmeans(multishapesData, 5, nstart = 20)
```

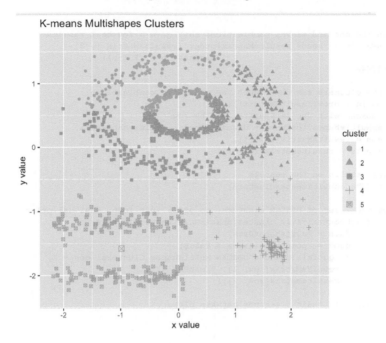

FIGURE 6.14: Clusters obtained in the Multishapes Dataset Using *K*-means

```
18
19  #Plot kmeans clusters
20  library (factoextra)
21  factoextra::fviz_cluster(kmeansMultishapesClusters, data =    multishapesData,
22          geom = "point", ellipse = FALSE,
23          main = "K-means Multishapes Clusters")
24
25  # Find suitable DBSCAN parameters: Perform grid search
26  library (c2c)
27  epsVals <- seq (0.1, 1, 0.05)
28  minPtsVals <- seq (4, 10)
29  resultsMatrix <- matrix (, ncol = 3)
30  #Loop over a number of values of eps and minPts, and keep track of
31  #the combination that gives maximum value of overall_purity using
32  #c2c library
33  for (epsVal in epsVals) {
34    for (minPtsVal in minPtsVals)
35      {
36        dbscanMultishapesClusters <- dbscan::dbscan (multishapesData,
37          eps = epsVal, minPts = minPtsVal )
38        confusionMatrix <- table(dbscanMultishapesClusters$cluster,
39                        multishapesLabels)
40        extrinsicAssess <- c2c::calculate_clustering_metrics(confusionMatrix)
41        resultsMatrix <- rbind (resultsMatrix, c(epsVal, minPtsVal,
42          extrinsicAssess$overall_purity))
43      }
44  }
```

```
45
46  #Obtain the eps and minPts combinations that give a few best results
47  resultsMatrix <- head( resultsMatrix[order (resultsMatrix [, 3],
48                                   decreasing = TRUE),])
49  ###########
50
51  #Plot the clusters for the "best" few parameter combinations
52  for (row in 1:nrow(resultsMatrix)){
53    #run dbscan with the parameters in  the row
54    dbscanMultishapesClusters <- dbscan::dbscan (multishapesData,
55        eps = resultsMatrix[row, 1],
56        minPts = resultsMatrix[row, 2])
57
58    #print details on the console
59    print (dbscanMultishapesClusters)
60
61    #Plot the  set of DBSCAN clusters, to print in a loop, must use print
62    print (factoextra::fviz_cluster(dbscanMultishapesClusters,
63              data = multishapesData,
64              geom = "point", ellipse = FALSE)
65              + ggtitle(paste ("DBSCAN Multishapes Clusters: eps = ",
66              resultsMatrix[row, 1], " minPts =",
67              resultsMatrix[row, 2]))
68          )
69    Sys.sleep (5)  #wait 5 seconds between plots
70  }
```

Lines 4-23 of this program downloads the `multishapes` dataset, clusters it using K-means, looking for 5 clusters, and plots the clusters as shown in Figure 6.14.

Lines 25-47 perform clustering of the dataset using DBSCAN. Lines 26-42 go through a loop where the program tries `eps` values between 0.1 and 1 with a jump of 0.05, and `minPts` values between 4 and 10 with a jump of 1. It runs `dbscan` for every combination of the two parameter values (lines 36-37), creates a confusion matrix (lines 38-39) and performs computation of extrinsic assessment As it loops, it keeps the `eps` and `minPts` values as well as overall purity metric in a matrix called `resultMatrix` (lines 29, and 41-42).

Line 47 sorts the `resultMatrix` by descending value of purity for combinations of *eps* and *minPts* values, and keeps only the top few rows as given by the `head` function. The best values for purity come from the parameter combinations given below.

```
>resultsMatrix
      [,1] [,2]       [,3]
[1,] 0.15    6 0.9763636
[2,] 0.20   10 0.9763636
[3,] 0.15    5 0.9754545
[4,] 0.15    4 0.9736364
[5,] 0.15    7 0.9736364
[6,] 0.15    8 0.9681818
```

Lines 61-70 loop through the rows of `resultsMatrix`, use the parameter values in each row to perform `dbscan` again (lines 54-56), print some details of

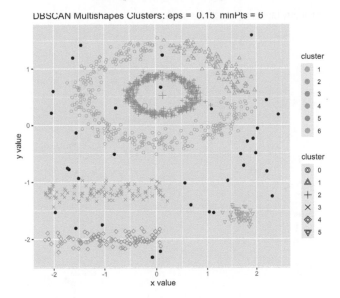

FIGURE 6.15: Clusters of Multishapes Dataset using tt eps=0.15 and minPts = 6

the clustering performed (line 59), and plot the set of clusters (lines 61-69). Since it takes a while to look at a plot carefully, the program waits 5 seconds (line 69) between running `dbscan` on the next set of parameter combination and plotting it. We find that the top ranked parameter combination of `eps` value of 0.15 and `minPts` value of 6 actually does not provide good clustering even though the purity value is the highest. It gives 6 clusters instead of the expected 5, breaking the outer ellipsoidal shape into two clusters, with the top right quadrant of it forming a cluster by itself. The second row, with the same purity value as the first row, with `eps` value of 0.20 and `minPts` value of 10 gives the right number of clusters with the right shapes. So do the combinations $\langle 0.15, 5 \rangle$, and $\langle 0.15, 4 \rangle$. The last two combinations produce 7 clusters each, and therefore are not correct. Thus, the point is that purity alone is not a good extrinsic measurement metric for clustering. In this case, we have domain knowledge, i.e., we know the shapes of the clusters, and therefore, could go through a list of top-ranked parameter combinations to find one or more combinations that provide clustering results that are consistent with domain knowledge. Clusters produced by $\langle 0.15, 6 \rangle$, and $\langle 0.20, 10 \rangle$ are given in Figure 6.15 and Figure 6.16, respectively. Thus, a combination of clustering metrics should be used to select clusters, not just one as this program does.

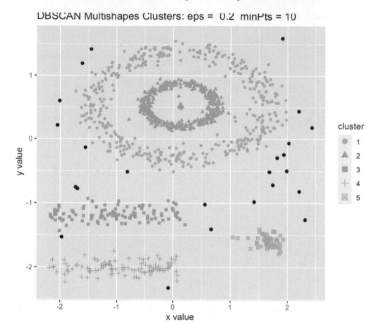

FIGURE 6.16: Clusters obtained in the Multishapes Dataset Using K-means

6.7   Conclusions

This chapter introduced the concept of unsupervised learning, in particular, the idea of clustering. It also presented three different types of clustering algorithms: centroid-based, hierarchical and density-based. For each of the types of clustering algorithms, it discussed how R libraries can be used to program clustering algorithms.

Exercises

1. Measuring distance between two items or a distance metric is essential for clustering. In this chapter, we discussed only one such metric, the Euclidean distance. Name a few other distance metrics that can be used in K-means type clustering. What are the advantages and disadvantages of the various distance metrics?

2. K-means clustering can be thought of as an optimization algorithm over an unlabeled dataset in order to identify subsets where each element in each subset are as similar as possible, and elements in different clusters are as dissimilar as possible. These requirements can be expressed in terms of a mathematical objective function for optimization. Formulate this objective function for a dataset with N unlabeled data points, each with n features or attributes.

3. The K-means algorithm is used to find K clusters in a dataset for a pre-specified value of K. How can K-means algorithm be used to find outliers in a dataset?

4. The clustering algorithms discussed in this chapter are examples of hard clustering in that the set of clusters is disjoint. In other words, each data example is allocated to one and only one cluster, and each data example is allocated to a cluster.

5. In hierarchical clustering, in applications in domains such as bioinformatics, the height of the dendograms is important. In particular, the height at which division of a bigger cluster to a smaller cluster happens is important. Why is it so? How can the distances be computed?

6. The way agglomerative clustering has been described in this book, the process starts with each data example in a cluster by itself. Then, it iterates finding a pair of clusters to merge till there is a single cluster at the end. Thus, the process is greedy in the sense that two clusters, once merged, cannot be "un-merged". Is this always good? If not, how can the process be improved?

7. Examples in some datasets are very high dimensional. In other words, the number of features is large, say in the thousands, or tens or thousands, or even more. For example, very high dimensional datasets occur in bioinformatics. How would one go about clustering very high-dimensional data?

8. There is an internet project called the *Tree of Life* that seeks to build a phylogeny tree for all of the living species in life, at least as many as possible. Discuss how it may be possible to cluster a large number of data examples, say a few thousand or hundreds of thousands or even millions.

9. In this book, we have discussed algorithms that can cluster only numerical data. In other words, all features or attributes of the data examples are numeric. However, in a real dataset, some or all of the attributes may be discrete. A discrete attribute takes a small number of distinct values. An example of a discrete attribute may be a logical attribute that takes only two values. Another example is the days of the week.

How can one cluster a dataset where each attribute is discrete? What about when some of the attributes are discrete and others numeric?

10. Clustering of a set or corpus of documents is often useful. Documents are made up of words, predominantly, and not numbers. How would you go about clustering a set of documents?

Chapter 7

Conclusions

This book has discussed several significant topics in machine learning in a manner suitable for advanced undergraduates and beginning graduate students in computer science and related fields. However, it an introductory book and as a result, many relevant topics have not been discussed and are left for the student to acquire on his/her own or by taking more advanced classes. To provide some guidelines for students who want to pursue further studies in machine learning, this chapter provides a list of valuable sources to pursue.

We recount the topics presented in the book one by one. Regression is a widely studied topic in machine learning and statistics. Chapter 2 of this book covered the fundamentals of linear least squares regression, its analytical derivation and ways to make it robust via regularization. A student who wants to study this topic further could either focus on the mathematical aspects of it, or the computational and practical aspects, or a combination of the two. Those mathematically minded can find many undergraduate and graduate level texts on the topic. It is a good idea to be conversant with both theoretical and practical perspectives. A highly recommended book is *Introduction to Linear Regression Analysis*, 5th edition, by Montgomery and Peck, 2021 [23]. It combines theoretical discussions with practical applications of regression analysis. Those who want to pursue regression from a machine learning perspective, but still understand the theory well may find Hastie, Tibshirani and Friedman's *Elements of Statistical Learning: Data Mining, Inference and Prediction*, 2nd Edition, 2008 [11] quite helpful. Chapter 3 of Hastie et al.'s book contains a very good description of theoretical as well as practical issues in regression as commonly used in machine learning. Hastie and Tibshirani [14] have also produced a very useful software library called *Generalized Additive Model (GAM)* [14] for performing simple as well as complex fitting of linear predictor models. The library was initially written in R, but is also available in Python.

Chapter 3 of the book presented tree-based classification models. Tree-based models are easy to understand and easy to interpret. To overcome shortcomings of individual tree learners and predictors, ensemble approaches such as random forests and boosting trees have been developed. Tree-based ensemble models are competitive, especially when the amount of data to train on is limited and/or computing resources are limited as well, and still obtain good results. In recent years, in many of the machine learning competitions organized by a company called Kaggle[1], the winner has been a regularized boosting tree-based approach called *Extreme Gradient Boosting (XGBOOST)*. XGBOOST is available for direct use as part of machine libraries in R and Python. The specific algorithm has not been discussed in this book because the associated concepts are a little advanced, but interested readers can read the paper from 2017 that describes its technical details [6].

Chapter 4 of the book introduces the basic concepts of artificial neural networks, and provides a glimpse of deep learning [13] in terms of discussion of convolutional neural networks (CNNs). Although CNNs were introduced by LeCun in 1998 [18], they started becoming popular in the machine learning community only in the early 2010s. CNN-based artificial neural networks first won a large-scaled image classification contest called ILSVRC (ImageNet Large Scale Visual Recognition Challenge)[2], besting the nearest competitor by more than 15% in accuracy. CNN-based ILSVRC contestants won the competition for the next several years also, establishing deep learning or deep artificial neural networks as the way to solve large-scale computer vision problems. Deep learning (whether CNN-based or otherwise) models are on the ImageNet leaderboard[3] since 2012, although some of the recent ones are not CNN-based, but are attention-powered, following a well-known architecture called the Transformer [32]. In this book, due to lack of space, we have not covered recurrent neural networks (RNNs) or advanced versions of RNNs such as LSTMs (Long Short-Term Memory) [17]. We also have not covered the so-called attention mechanism [] and complex modular string(language)-processing approaches such as Transformers introduced in 2017, and even more recent approaches BERT (Bidirectional Transformers for Language Understanding) [7]. Transformer-based approaches such as BERT have become dominant in natural language processing (NLP), an area focused on building tools for human languages in contrast to machine languages (programming languages). For neural networks to process human languages well, words in a language have to be transformed into a numeric form. There are techniques called Word2Vec [19], Glove [24] and BERT-encoding [7] that take into account how words co-occur in a large body (also called corpus) of text so that textual documents can be processed by neural networks. Deep learning is a dynamic and evolving field of study with new approaches discovered at a fast clip. The interested student is advised to look at published cutting-edge papers in well-known conferences such as NeurIPS (Neural Information Processing Systems)[4], ICML (International Conference on Machine Learning)[5], ICLR (International Conference on Learning Representations)[6],

---

[1]kaggle.com
[2]https://image-net.org/challenges/LSVRC/
[3]https://paperswithcode.com/sota/image-classification-on-imagenet
[4]https://nips.cc
[5]https://icml.cc
[6]https://iclr.cc

AAAI (Association for Advancement of Artificial Intelligence)[7], ACL (Association for Computational Linguistics)[8], and NAACL-HLT (North American Association for Computational Linguistics—Human Language Technologies)[9]. One should also look at well-known journals like *IEEE Transactions on Pattern Analysis and Machine Intelligence, IEEE Transactions on Neural Networks and Learning Systems,* and Elsevier's *Neural Networks.*

Chapter 5 of the book covers another important topic called reinforcement learning. Unlike supervised and unsupervised learning, the goal of reinforcement learning is to learn an optimal policy that informs an agent what action to perform in which state so as to maximize the cumulative rewards received. The definitive book on this topic is *Reinforcement Learning: An Introduction,* Second Edition, 2018, by Sutton and Barto [31]. Although it is an updated recent version of an older edition, it covers mostly the fundamentals of reinforcement learning, without much emphasis on deep reinforcement learning. Reinforcement learning is used to learn a function that outputs the most appropriate action, given the current state and maybe, one or more past states, as input. If the input size is large, e.g., a high-fidelity frame of a video game, exact algorithms are unable to learn the state-to-action mapping well, if it is at all possible. In such cases, learning the function approximately can be achieved via the use of deep artificial neural networks. *Spinning Up in Deep RL!*[10] by OpenAI is a modern introduction to reinforcement learning. It covers the essential theory and then provides guidance to implementing sophisticated deep learning algorithms quickly in a software environment called *Gym.* Mnih et al.'s 2013 paper [21] in NeurIPS (followed by a 2015 paper in *Nature* [22]) on developing reinforcement learning agent programs to play human level Atari games, and Silver et al.'s 2016 *Nature* paper on learning to play champion-level Go [30], firmly placed deep learning as a very competent modern software tools to solve many task-oriented machine learning problems. Papers on recent developments on deep reinforcement learning are found in mainstream machine learning conferences and journals mentioned earlier.

Clustering, the topic of Chapter 6, is an exploratory data analysis technique. This book discussed centroid-based, hierarchical and density-based clustering. Due to lack of space, we just scratched the surface in exploring algorithms and issues in clustering. A student who wants to explore this topic further may start with a book like *Data Clustering: Algorithms and Applications,* 2014, by Agarwal and Reddy [1]. It covers topics such as probabilistic and spectral clustering, stream and time-series data clustering, uncertain data clustering, and applications of clustering in many areas. Latest developments in clustering and exploratory data analysis are published in KDD (ACM SIGKDD Conference on Knowledge Discovery and Data Mining)[11],

---

[7]aaai.org
[8]https://www.aclweb.org/anthology/
[9]https://www.aclweb.org/anthology/
[10]https://spinningup.openai.com/en/latest/
[11]https://kdd.org/conferences

ICDE (International Conference on Data Engineering)[12], and CIKM (International Conference on Information and Knowledge Management)[13].

To summarize, this book has been prepared to contain enough material for a one-semester introductory course on machine learning. As a result, a limited number of topics were chosen expressly for the purpose. We omitted a number of topics on purpose. However, planning ahead for a second edition, we would like to add a section of topics such as K-nearest neighbors classification, logistic regression, support vector machines (SVMs) and naive Bayes classifier. These are commonly used classification algorithms, especially when the amount of data is not large and the data are expressed in terms of features. We would like to add a discussion of recurrent neural networks as well. In the supervised learning chapter, we would like to add discussions on a selection of topics such as association rule mining, outlier mining, feature selection and creation of word embeddings so that text can be processed using neural networks. Since the addition of these topics will increase the number of pages in the book beyond the 250-300 pages that can be realistically covered in a semester, the current plan for the second edition is to move all the code to the companion website and keep only the theoretical discussions in the book.

---

[12]http://ieee-icde.org
[13]http://www.cikmconference.org

Bibliography

[1] Charu C Aggarwal and Chandan K Reddy. Data clustering. *Algorithms and applications. Chapman&Hall/CRC Data mining and Knowledge Discovery series, Londra*, 2014.

[2] Esteban Alfaro, Matías Gámez, Noelia Garcia, et al. Adabag: An r package for classification with boosting and bagging. *Journal of Statistical Software*, 54(2):1–35, 2013.

[3] L Breiman and A Cutler. Setting up, using, and understanding random forests v4. 0. *University of California, Department of Statistics*, 2003.

[4] Leo Breiman. Random forests. *Machine learning*, 45(1):5–32, 2001.

[5] Peter Brown, H Roediger, Mark McDaniel, and Make It Stick. The science of successful learning. *Cambridge, MA*, 2014.

[6] Tianqi Chen and Carlos Guestrin. Xgboost: A scalable tree boosting system. In *Proceedings of the 22nd ACM SIGKDD International Conference on Knowledge Discovery and Data Mining*, pages 785–794, 2016.

[7] Jacob Devlin, Ming-Wei Chang, Kenton Lee, and Kristina Toutanova. Bert: Pre-training of deep bidirectional transformers for language understanding. *arXiv preprint arXiv:1810.04805*, 2018.

[8] Thomas G Dietterich. Ensemble methods in machine learning. In *International workshop on multiple classifier systems*, pages 1–15. Springer, 2000.

[9] Yoav Freund, Robert Schapire, and Naoki Abe. A short introduction to boosting. *Journal-Japanese Society for Artificial Intelligence*, 14(771-780):1612, 1999.

[10] Yoav Freund, Robert E Schapire, et al. *Experiments with a new boosting algorithm*. In *ICML*, volume 96, pages 148–156. Citeseer, 1996.

[11] Jerome Friedman, Trevor Hastie, and Robert Tibshirani. *The elements of statistical learning*, volume 1. Springer series in statistics New York, 2001, 2008.

[12] WF Giauque and R Wiebe. The heat capacity of hydrogen bromide from 15 k. to its boiling point and its heat of vaporization. the entropy from spectroscopic data. *Journal of the American Chemical Society*, 50(8):2193–2202, 1928.

[13] Ian Goodfellow, Yoshua Bengio, Aaron Courville, and Yoshua Bengio. *Deep learning*, volume 1. MIT press Cambridge, 2016.

[14] Trevor Hastie and Robert Tibshirani. Generalized additive models: some applications. *Journal of the American Statistical Association*, 82(398):371–386, 1987.

[15] Robert Higgs. Race, skills, and earnings: American immigrants in 1909. *The Journal of Economic History*, 31(2):420–428, 1971.

[16] Tin Kam Ho. Random decision forests. In *Proceedings of 3rd international conference on document analysis and recognition*, volume 1, pages 278–282. IEEE, 1995.

[17] Sepp Hochreiter and Jürgen Schmidhuber. Long short-term memory. *Neural computation*, 9(8):1735–1780, 1997.

[18] Yann LeCun, Léon Bottou, Yoshua Bengio, and Patrick Haffner. Gradient-based learning applied to document recognition. *Proceedings of the IEEE*, 86(11):2278–2324, 1998.

[19] Tomas Mikolov, Kai Chen, Greg Corrado, and Jeffrey Dean. Efficient estimation of word representations in vector space. *arXiv preprint arXiv:1301.3781*, 2013.

[20] Thomas Mitchell. Machine learning, McGraw-Hill higher education. *New York*, 1997.

[21] Volodymyr Mnih, Koray Kavukcuoglu, David Silver, Alex Graves, Ioannis Antonoglou, Daan Wierstra, and Martin Riedmiller. Playing atari with deep reinforcement learning. *Neural Information Processing*, 2013.

[22] Volodymyr Mnih, Koray Kavukcuoglu, David Silver, Andrei A Rusu, Joel Veness, Marc G Bellemare, Alex Graves, Martin Riedmiller, Andreas K Fidjeland, Georg Ostrovski, et al. Human-level control through deep reinforcement learning. *nature*, 518(7540):529–533, 2015.

[23] Douglas C. Montgomery, Elizabeth A. Peck, and G. Geoffrey Vining. *Introduction to Linear Regression Analysis*. John Wiley & Sons, 2021.

[24] Jeffrey Pennington, Richard Socher, and Christopher D Manning. Glove: Global vectors for word representation. In *Proceedings of the 2014 conference on empirical methods in natural language processing (EMNLP)*, pages 1532–1543, 2014.

[25] J Ross Quinlan. *C4. 5: programs for machine learning*. Elsevier, 1993.

[26] R Quinlan. C5. 0: An informal tutorial. rulequest, 1998, rulequest.com, 1998.

[27] Gross Richard. Psychology the science of mind and behavior, 2009.

[28] Raúl Rojas. Adaboost and the super bowl of classifiers a tutorial introduction to adaptive boosting. *Freie University, Berlin, Tech. Rep*, 2009.

[29] Arthur L. Samuel. Some studies in machine learning using the game of checkers. *IBM Journal of research and development*, 3(3):210–229, 1959.

[30] David Silver, Aja Huang, Chris J Maddison, Arthur Guez, Laurent Sifre, George Van Den Driessche, Julian Schrittwieser, Ioannis Antonoglou, Veda Panneershelvam, Marc Lanctot, et al. Mastering the game of go with deep neural networks and tree search. *nature*, 529(7587):484–489, 2016.

[31] Richard S. Sutton and Andrew G. Barto. *Reinforcement learning: An introduction*. MIT press, 2018.

[32] Ashish Vaswani, Noam Shazeer, Niki Parmar, Jakob Uszkoreit, Llion Jones, Aidan N Gomez, Lukasz Kaiser, and Illia Polosukhin. Attention is all you need. *arXiv preprint arXiv:1706.03762*, 2017.

[33] John G. Wagner, George K. Aghajanian, and Oscar HL Bing. Correlation of performance test scores with "tissue concentration" of lysergic acid diethylamide in human subjects. *Clinical Pharmacology & Therapeutics*, 9(5):635–638, 1968.

Index